Zinc Surfaces

ZAHNER'S ARCHITECTURAL METALS SERIES

Zahner's Architectural Metals Series offers in-depth coverage of metals used in architecture and art today. Metals in architecture are selected for their durability, strength, and resistance to weather. The metals covered in this series are used extensively in the built environments that make up our world and are also finding appeal and fascination to the artist. These heavily illustrated guides offer comprehensive coverage of how each metal is used in creating surfaces for building exteriors, interiors, and art sculpture. This series provides architects, metal fabricators and developers, design professionals, and students in architecture and design programs with a logical framework for the selection and use of metallic building materials. Forthcoming books in Zahner's Architectural Metals Series will include Copper, Brass, and Bronze; Steel; and Zinc surfaces.

Titles in Zahner's Architectural Metals Series include:

Stainless Steel Surfaces: A Guide to Alloys, Finishes, Fabrication, and Maintenance in Architecture and Art

Aluminum Surfaces: A Guide to Alloys, Finishes, Fabrication, and Maintenance in Architecture and Art

Copper, Brass, and Bronze Surfaces: A Guide to Alloys, Finishes, Fabrication, and Maintenance in Architecture and Art

Steel Surfaces: A Guide to Alloys, Finishes, Fabrication, and Maintenance in Architecture and Art

Zinc Surfaces: A Guide to Alloys, Finishes, Fabrication, and Maintenance in Architecture and Art

Zinc Surfaces

A Guide to Alloys, Finishes, Fabrication, and Maintenance in Architecture and Art

L. William Zahner

WILEY

This book is printed on acid-free paper.

Copyright © 2021 by John Wiley & Sons, Inc. All rights reserved

Published by John Wiley & Sons, Inc., Hoboken, New Jersey
Published simultaneously in Canada

No part of this publication may be reproduced, stored in a retrieval system, or transmitted in any form or by any means, electronic, mechanical, photocopying, recording, scanning, or otherwise, except as permitted under Section 107 or 108 of the 1976 United States Copyright Act, without either the prior written permission of the Publisher, or authorization through payment of the appropriate per-copy fee to the Copyright Clearance Center, 222 Rosewood Drive, Danvers, MA 01923, (978) 750-8400, fax (978) 646-8600, or on the web at www.copyright.com. Requests to the Publisher for permission should be addressed to the Permissions Department, John Wiley & Sons, Inc., 111 River Street, Hoboken, NJ 07030, (201) 748-6011, fax (201) 748-6008, or online at www.wiley.com/go/permissions.

Limit of Liability/Disclaimer of Warranty:

While the publisher and author have used their best efforts in preparing this book, they make no representations or warranties with the respect to the accuracy or completeness of the contents of this book and specifically disclaim any implied warranties of merchantability or fitness for a particular purpose. No warranty may be created or extended by sales representatives or written sales materials. The advice and strategies contained herein may not be suitable for your situation. You should consult with a professional where appropriate. Neither the publisher nor the author shall be liable for damages arising herefrom.

For general information about our other products and services, please contact our Customer Care Department within the United States at (800) 762-2974, outside the United States at (317) 572-3993 or fax (317) 572-4002.

Wiley publishes in a variety of print and electronic formats and by print-on-demand. Some material included with standard print versions of this book may not be included in ebooks or in print-on-demand. If this book refers to media such as a CD or DVD that is not included in the version you purchased, you may download this material at http://booksupport.wiley.com. For more information about Wiley products, visit www.wiley.com.

Library of Congress Cataloging-in-Publication Data
Names: Zahner, L. William, author. | John Wiley & Sons, publisher.
Title: Zinc surfaces: a guide to alloys, finishes, fabrication, and
   maintenance in architecture and art / L. William Zahner.
Other titles: Zahner's architectural metals series
Description: Hoboken, New Jersey : Wiley, [2021] | Series: Zahner's
   Architectural Metals Series
Identifiers: LCCN 2021003278 (print) | LCCN 2021003279 (ebook) | ISBN
   9781119541615 (paperback) | ISBN 9781119541639 (adobe pdf) | ISBN
   9781119541592 (epub)
Subjects: LCSH: Zinc—Surfaces. | Zinc—Finishing. | Zinc coatings. |
   Architectural metal-work. | Art metal-work.
Classification: LCC TS640 .Z34 2021 (print) | LCC TS640 (ebook) | DDC
   661/.0661—dc23
LC record available at https://lccn.loc.gov/2021003278
LC ebook record available at https://lccn.loc.gov/2021003279

Cover Design: Wiley
Cover Images: (Main) *Steven Holl Architects', Institute for Contemporary Art (ICA), Virginia Commonwealth University*, photographed by Iwan Bann, c. 2017. (Border) *Pattern on zinc plates* © somnuk / Getty Images

This book is dedicated to my good friend Verne Christensen.
(a designer who proved Alessandro Volta's theory when we built a beautiful curved zinc roof below his copper gutters)

Contents

	Preface	xiii
CHAPTER 1	**Introduction to Zinc**	**1**
	Element 30 Zn – *Spelter*	1
	The Zinc Atom	9
	History	11
	Zinc Mineral Forms	17
	Zinc in Art	18
	Zinc as an Architectural Metal	24
	Health and Hygiene	28
	The Enigmatic Metal	31
CHAPTER 2	**Zinc Alloys**	**33**
	Introduction	33
	Alloying Descriptions	34
	Ingot Alloys	36
	Zinc Alloys – Rolled Forms	39
	Zinc Alloys Used in Architecture	41
	Wrought Zinc Alloys	44
	Architectural Rolled Zinc	47
	Forged and Extruded Zinc Alloys	53
	Cast Zinc Alloys	55
	Slush Casting	56
	Zinc Die Casting	58
	Gravity Cast Alloys	59
	Kirksite	62
CHAPTER 3	**Finishes**	**63**
	Introduction	63
	Appearance among Metals	65
	Mill Finishes	68
	Natural Zinc Color	68
	Mechanical Finishes	71
	Mechanically Rolled Textures	72

	Preweathered Zinc Surface	73
	Clear Coating with Pigmentation	77
	Blackened Zinc	77
	Custom Patina Finish	79
	Dark Variegated Patinas on Zinc	80
	Zinc Oxide Patinas	87
	Zinc Iridescent Patina	93
	Galvanized Zinc Surfaces	93
	Galvanized Steel Structural Shapes	99
	Darkening Galvanized Steel	100
	Zinc Phosphate Coatings on Galvanized Steel	101
	Zinc Fabric	102
	Other Methods of Applying Zinc to Steel	103
	Zinc Anodizing	104
CHAPTER 4	Expectations	105
	Introduction	105
	Natural Finish on Thin Sheet Material	107
	Natural Finish on Thick Plate Material	109
	Natural Finish on Cast Surface	110
	Preweathered Finish	113
	Preweathered with Added Pigmentation	119
	Expectations – Preweathered Surface	120
	Blackened Zinc	121
	Color Matching	122
	Custom Patinas	124
	Flatness and Visual Distortion	131
	Creep	135
	Galvanized Surface	138
	Darkened Galvanized Steel	142
CHAPTER 5	Available Forms	145
	Introduction	145
	Wrought Forms of Zinc	148
	Plate	152
	Sheet and Coil	153
	Zinc Foil	158
	Extrusion	158
	Tube and Pipe	159
	Wire	160
	Rod	160
	Wire Mesh	160
	Expanded Metal	160

	Perforated Zinc	162
	Textured Zinc Sheet	163
	Zinc Ornamentation	165
	Cast	166
	Slush Cast	167
	Die Cast	167
	Sand Cast	169
	Zinc Powder	171
CHAPTER 6	Fabrication	173
	Working with Zinc	173
	Storage and Handling	174
	Cutting Zinc	177
	Shearing and Blanking	177
	Saw Cutting	177
	Laser	178
	Plasma	179
	Waterjet	179
	Punching / Perforating / Bumping	180
	Forming and Bending	184
	Grain Direction and Anisotropy	185
	Temperature Effect on Forming	187
	Brake Forming	187
	V-Cutting	191
	Roll Forming	191
	Superplastic Forming	192
	Forging	193
	Extrusion	194
	Machining	194
	Soldering	195
	Welding	197
	Fusion Stud Welding	199
	Resistant Welding of Zinc	201
	Expansion / Contraction	202
	Bolting and Fastening	204
	Thermal Spray	206
	Hot-Dipped Galvanizing	206
	Casting	208
	Die Casting	210
	Slush Casting	211
	Permanent Mold Casting	212
	Sand Casting	212
	Plaster Mold Casting	212
	Spin Casting	212

CHAPTER 7 Corrosion — 215

- Introduction — 215
- Zinc as a Protective Coating — 216
- Galvanized Steel — 218
- Zinc Alloy Coatings on Steel — 219
- Zinc Powder in Paint Coatings — 220
- Sherardizing — 221
- Thermal Spray — 221
- Zinc Anodes — 222
- Battery — 222
- When Zinc Does Not Protect Steel — 224
- Zinc Corrosion — 225
- Interior Exposures — 227
- Exterior Exposures — 228
- Sheltered Exterior Surfaces — 230
- Uniform Corrosion — 235
- Underside Corrosion — 236
- Wet Storage Stain — 237
- Galvanic Corrosion — 239
- Determining Factors for Galvanic Corrosion — 242
- Difference in Electro-Potential — 243
- Geometric Relationship — 243
- Distance — 244
- Electrolyte Effects — 244
- Temperature Effects — 245
- Pitting Corrosion — 246
- Intergranular Corrosion — 248
- Stress Corrosion Cracking — 248
- Zinc Artifacts and Statues — 249
- Deicing Salts — 251
- Chlorides — 252
- Fertilizer — 253
- Saponification — 254
- Corrosive Substances in Proximity — 254

CHAPTER 8 Maintaining the Zinc Surface — 257

- Introduction — 257
- Zinc Surfaces — 258
- Why a Maintenance Procedure — 260
- Develop a Maintenance Strategy — 260
- Restoring the Preweathered Appearance — 264
- Effects of Different Environments — 266

	Physical Cleanliness	267
	Chemical Cleanliness	278
	Mechanical Cleanliness	292
	Galvanized Steel Surfaces	296
APPENDIX A	Brand Names	301
APPENDIX B	Select Specifications for Zinc	303
	References	305
	Index	307

Preface

An expert is an ordinary man who- when he is not home – gives advice.

Oscar Wilde

Zinc is the mysterious metal used in art and architecture.

In the United States, it is a paradox. On the one hand, it is considered an Old World metal, used for centuries across Europe. Paris is defined by the roofs of zinc that blanket the city. Yet it is relatively new to North American architecture.

As a metal of art casting or fenestration, little was known until the early 1990s. Sure, we knew of the process of dipping steel in a molten jacket of zinc. Hot-dipped galvanized, a strange fondue for metal, is a process that is well known, but not always understood.

The leading zinc mines that supplied the world were once in the middle of the United States, a region with the town of Joplin, Missouri, as the center. Most zinc mining has ended in the area, but in the late 1800s and early 1900s this was the epicenter of zinc. Millionaires were made by the dozens as the area was tunneled out like a giant anthill.

In North America, the sheet metal industry, art casting industry, and design community knew little about zinc. Publications and training documents throughout the architectural metal industry made no mention of zinc. The old catalogs called the metal *white bronze,* perhaps attempting to elicit a feeling of noteworthiness by taking on the name *bronze*. Metal foundries, art schools, and metal workers in the United States lacked any real knowledge of the metal. With the exception of galvanizing, the metal was all but forgotten after the early part of the twentieth century.

When I first started work at Zahner, a 125-year-old metal fabrication company located in the Midwest, zinc was not known as an architectural metal. We did not stock the metal, nor was it specified in any industry publications. We worked with steel, terne, copper, aluminum, stainless steel, and lead, but not zinc. One of the first introductions to the metal occurred during the restoration of the Folly Theater, a turn-of-the-century theater built in 1900. When the workers removed parts of the metal cornice and decorative metal baluster in 1979, they had difficulty determining what the metal was. This metal had lasted 80 years and still looked in good shape. It was silver under the paint, so it was not copper. It was not magnetic so it was not terne-plated steel or galvanized steel. It was heavy, so it wasn't aluminum, and aluminum had not been in common use by 1900. The pieces were spun and assembled in sections by soldering. It was not any metal we were familiar with. It was zinc. From the *old country*.

The other connection to zinc goes back 125 years. Andrew Zahner, my great grandfather, started this metal company I work for, in Joplin, Missouri, in 1897. Back in the late 1800s, this region in southwestern Missouri, on the edge of the Ozark Mountains, was the site of one of the largest mining operations in America – first for lead used to make bullets and later for zinc. Zinc, known in the area as *jack,* made the region one of the wealthiest in the United States. Every major railroad at the time went through the Joplin region to transport the ore around the United States and to ports to supply the European market. The ore was of such high quality that the Europeans purchased it from Missouri.

This booming region attracted a young Andrew Zahner, and he started a small metal fabrication firm to produce cornices and other decorative features for the wealthy merchants in the area.

Andrew Zahner surely knew about zinc.

The boom / bust cycle eventually hit the Joplin area in the early 1900s, and Andrew moved the company to its current home in Kansas City. The knowledge of zinc was left behind with the dying mines of the central United States. Now, 125 years later, I write a book on this metal, zinc. It is unfortunate that I did not have Andrew as a resource.

Over the last couple of decades, we have worked with the metal zinc on numerous projects. We have expanded our knowledge of the metal and have uncovered many new and interesting ways of working with zinc. We have created new patinas and surface enhancements, and we have explored casting. The more I work with the metal zinc, the more I find it an intriguing material of design.

Working with my daughter Kat, who operates Zahner Metal Conservation, restoring 100-year-old zinc statues and statuettes gives a deep appreciation for how the artist worked with the metal and produced amazing detail using casting techniques that have all been forgotten.

This book, the fifth in the series on metals, is intended to spark the interest in the metal zinc and explore the possibilities it has to offer the designer and the artist. The next pages should help to unravel this interesting material of design and introduce the reader to how this metal will appear and function.

L. William Zahner

CHAPTER 1

Introduction to Zinc

It's a business. If I could make more money down in the zinc mines, I'd be mining zinc.

Source: Roger Maris

ELEMENT 30 ZN – *SPELTER*

Zinc, the metal that could change copper into gold, at least that was the wish of the early alchemists. They called the metal *counterfeht*[1]. It looked like silver, but it wasn't. Adding it to molten copper and the copper would turn to a beautiful golden color, but it was not gold. It was an "imitation" a *counterfeht*. This odd metal, if it was a metal at all, was a mystery.

Zinc went also by the name *spelter,* used mainly by those who worked with the metal. Spelter was possibly a corruption of the name for "pewter," the dull gray, lead-tin alloy. The Dutch, first to import the metal into Europe used the word, *spiauter* for a word to describe a mixture of lead and tin[2]. So, it very well may have been an early marketing ploy to give value to this dubious metal. Spelter was the name given to this metal up until relatively recent times. Today, the name zinc has firmly taken hold on the periodic table of elements.

Other names, in particular *calamine,* were frequently used for this metal before it was officially a metal. Calamine, the principal mineral of zinc, was the name used across Western civilizations since the time of the Romans. Calamine is zinc sulfide, and there are regions in Europe where the rich mineral deposits of zinc sulfide were mined.

Calamine, as well, lost out as a name for the principal mineral form of zinc and is now better known as the popular topical poison ivy cream, even though the lotion contains zinc oxide, not the

---

[1] The German, Georgius Agricola, in his book, *De re metallica*, written in 1556.
[2] Dawkins J.M., *Zinc and Spelter,* 1950, Zinc Development Association, p. 24.

zinc sulfide of the mineral form. Instead of calamine, the term *sphalerite* is used as the name for the zinc sulfide mineral. For a long while the term *zinc blende,* from the German *zincblende,* was also used to describe the mineral. Confusion reigned on what this mineral or metal actually was.

As a metal, zinc in a wrought or cast form came late, sometime in the middle of the sixteenth century to the Western civilizations, definitely earlier in India. China also was an early zinc producer, using crucibles with charcoal to heat the ore. They made coins from zinc in the fourteenth century. The Romans would produce brass from copper by adding calamine and heating it in small crucibles. The zinc was obtained by reducing the ore, releasing carbon dioxide, and the fumes of zinc would rapidly be absorbed into the copper. Once melted, the slag-covered block would be hammered and the bright yellow color would appear.

The process of making brass was well known throughout antiquity. The method of creating brass from sphalerite (or calamine, as it was then known) was described in several texts. One such text, *Schedula Diversarium Artium*, written by Theophilus Presbyter in the eleventh century, describes the heating of crucibles in an open furnace, adding calamine, then strips of copper. Place back on the furnace for 9 hours and you arrive at a golden yellow color pleasing to look at. Figure 1.1 shows the mineral sphalerite with a large lump of copper on a plate of brass.

Zinc appeared as a known metal later than lead and tin. The mineral was known but as a distinct metal, zinc was not. Along with other colorful zinc minerals, sphalerite was easy to identify and so was mined in antiquity as a mineral to add to copper to produce the beautiful yellow brass. Granted, it was often mistaken for galena, a lead sulfide mineral, at one time a valuable mineral for making bullets.

Early brass artifacts dating back to the eighth century BCE were uncovered in the Gordion tomb excavated in Anatolia. The copper–zinc metal was called *oreichalkos* and later *orichalcum* by the Romans. The process of producing brass was well known and documented. Most brass production was established near the zinc mines because it was easier to cart copper to the area, than the large quantity of zinc mineral needed.

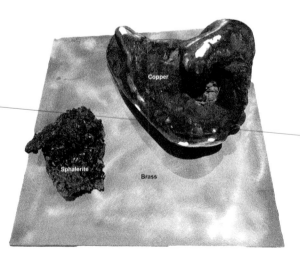

FIGURE 1.1 Sphalerite plus copper to make brass.

The reason zinc as a known metal was late to discovery is the difficulty of refinement. Up until the mid-1700s, metals were made by roasting the ores and burning off the oxides to free the metal. Trouble is, zinc has a low boiling point as metals go. As sufficient heat is applied to reduce the ore, zinc turns to gas and the fumes escape. Thus, the reduction of the ore the way other metals are produced just did not work for zinc.

The way the early alchemists found this *counterfeht*, it would condense on the walls of the flue and in cracks and crevices of the stone after roasting metal ores that contained zinc. Zinc is often found in ores of other metals, particularly lead, copper and silver. When the ores were heated the zinc would go up as vapors and condense on the stone. When it condensed, it formed long, whiskery tuffs the alchemists called *lana philosophica,* meaning "philosopher's wool."

Assistants to the alchemists would scrape and collect this wooly substance off the stone and out of the cracks of the flue walls. The alchemists placed a value on this special metal that was like tin but when added to copper would transform the copper into a golden yellow.

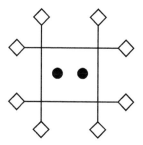

Alchemist symbol for zinc.

Zinc is a silver metal with a slight bluish hue. Zinc can be polished to a bright, silver but quickly tarnishes when handled. As zinc ages it turns to a rich gray color with whitish oxides in areas where moisture is allowed to accumulate.

Zinc is element 30 on the periodic table of elements, Figure 1.2. With the red metal, copper, on one side and gallium, a blue gray metal that melts in your hand, on the other, zinc falls in the twelfth row with cadmium and mercury.

Zinc has several isotopes, but the isotope zinc 67 is rather special. Zinc 67 occurs in approximately 4% of natural zinc. This isotope is highly sensitive to minute variations in transmitted energy. When it detects energy, it emits electromagnetic radiation making this isotope zinc 67 valuable for high accuracy measuring equipment. Zinc 67 is used to detect gamma ray vibrations with incredible sensitivity in the highly accurate atomic clock.

Zinc has a hexagonal crystal structure, which even though it is closely packed, it is less dense than the cubic structure of iron or copper. Figure 1.3. depicts a closely packed hexagonal crystal.

This metallurgical structure shows the crystal of zinc has six atoms in a near plane and another six slightly further away. This makes the bonds of the basal plane slightly stronger than the bonds of the parallel plane. This difference in distance and strength gives zinc an anisotropy that translates to forms made of zinc.[3]

---

[3] Porter, Frank, *Zinc Handbook*, Marcel Decker Inc., NY, p. 45.

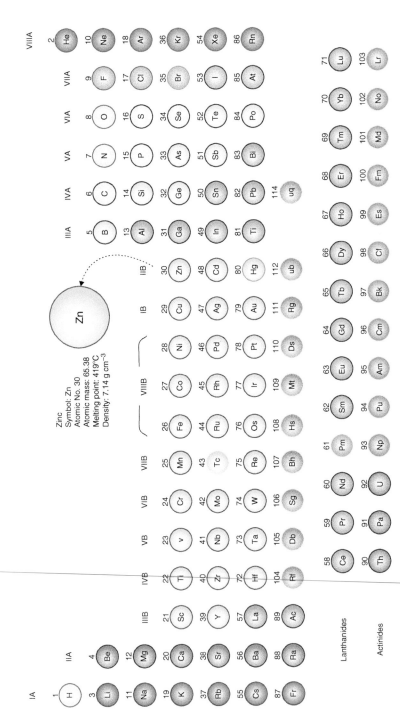

FIGURE 1.2 Periodic table.

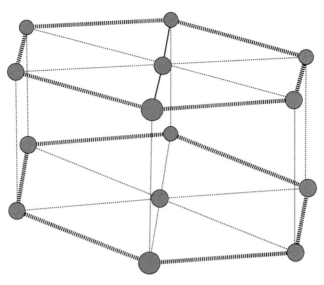

FIGURE 1.3 Zinc hexagonal crystal structure.

Another aspect of zinc is its ability to recrystallize rapidly after deformation. This prevents work hardening from occurring during forming operations and it also provides a level of "self-lubrication" as the crystals slip over one another during forming processes.

Note, the spangle that forms on galvanized steel is a large crystal of zinc that forms as it cools. It has the six triangular wedge-like symmetry reflecting the hexagonal crystal structure of the zinc crystal lattice. Figure 1.4 shows a close-up image of the spangle formed when zinc cools on a steel

FIGURE 1.4 Spangle of galvanized.

FIGURE 1.5 Zinc coating on steel using a controlled refinement of the cooling process.

substrate. The wedges that expand out from a central point are called dendrites and the parallel lines are called subdendrites. When newly developed the galvanized surface has a crystalline reflective quality due to the way the subdendrites scatter the reflective light. The surface seems to come alive as you walk around a newly galvanized steel plate with the glittering reflection bouncing off the variations in the crystals.

This reflectiveness, achieved by hot-dipping steel into molten zinc, is a natural surface that forms due to slight imperfections in the zinc bath or slight roughness on the steel surface. These imperfections initiate the formation of the dendrite growth.

Artistic affects can be enhanced to take advantage of cooling rates of the molten zinc. These techniques are still in development in order to better understanding the parameters involved. However, cooling rates, "seeding" the molten bath with other elements can influence the effects.

The difficulty arises in the industrial controls in place by the galvanizing facilities. Artistic expression is not in their normal parlance.

Figure 1.5 shows a "wave-like" appearance that has developed on flat steel sheet. The reflectivity enhances the three-dimensional appearance of the zinc surface.

As the surface oxidizes, the zinc crystals still vary in appearance creating a dull, lower reflective patchwork appearance. The dendrites are still there, they have just developed a layer of zinc oxide that mutes the reflectivity. Figure 1.6 shows a galvanized plate that has been exposed to weather.

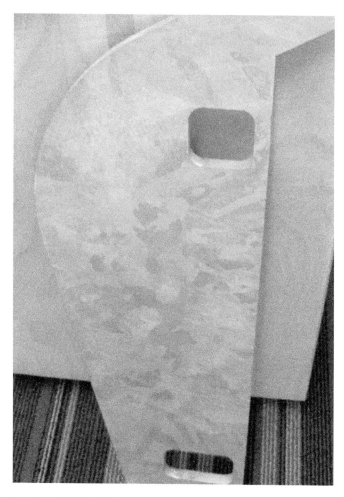

FIGURE 1.6 Galvanized that has weathered.

Chapter 1   Introduction to Zinc

Zinc	ELEMENT 30
Atomic number 30	
Crystal structure:	Close-packed hexagonal
Main mineral source:	Sphalerite (Calamine)
Color:	Bluish white
Oxide:	White
Density:	7,068 kg/m^3
Specific gravity:	7.0
Melting point:	419°C
Thermal conductivity:	112 W/m °C
Coefficient of linear expansion:	19×10^{-6} m/m°C
Electrical conductivity:	26% IACS
Modulus of elasticity:	108 GPa

Most of the zinc found on the Earth's surface is from hydrothermal activity that brought the metal to or near the surface. Zinc is not found in the native state. Zinc is always found in combination with other elements and metals. Zinc is the 24th most abundant element within the upper crust of the Earth.

Zinc has a poor strength to weight ratio as compared to other metals used in industry.

Zinc alloys are ductile at room temperature. Zinc castings are not ductile.

Zinc is subject to fracture when formed at low temperatures.

High elasticity – resiliency under shock loading

Soft edge

Zinc and zinc oxides are nontoxic unless consumed in large amounts. Zinc oxide fumes are hazardous when inhaled and will cause flu-like symptoms that can last 1–2 days.

It has superior corrosion resistance in many natural environments. Zinc is subject to corrosion in low pH and high pH environments.

FINISHES:

Mill finish – as rolled.

Semi-bright

Preweathered – darkened

Zinc can be painted.

Coil-coated zinc sheet in various colors are available on the market.

Oil-based paints are not recommended. Saponification can develop.

Plating with other metals such as copper, silver, nickel, and gold are possible.

Artificial patina:	Zinc can receive artificial patinas of white, black, browns, mottled browns with green and reddish oxides as well as iredescent hues of transparent greens, purples and reds.
Bright appearance:	Zinc can be polished but the luster quickly diminishes as oxides form. The color is typically a matte gray to grayish blue.
Reflectance of ultraviolet: of infrared:	The oxide of zinc absorbs ultraviolet light. Its use in sun protection is well known. Protection is afforded by absorption of the ultraviolet radiation and not allowing it to pass to the skin. Zinc oxide in powder form is used extensively in paint. It is a white powder and will reflect infrared radiation.
Relative cost:	Medium
Strengthening:	Zinc does not gain strength from cold working as other metals do. Instead, alloying with small amounts of copper and titanium are used to improve strength and add creep resistance.
Recycle ability:	Zinc is easily recycled because of the low melting point. Zinc is captured in from galvanize coated steels as vapor during the recycling process of coated steels.
Welding and joining:	Zinc can be welded and soldered.
Casting:	Zinc is a common casting metal. Used for many small cast parts where strength is less a requirement
Plating:	Zinc can be electroplated with other metals.
Etching and milling:	Zinc can be etched chemically and readily machined.

THE ZINC ATOM

All metals have at most three electrons in their outer shell. Zinc, element 30 on the periodic chart, has two electrons in the outer shell. This gives it an oxidation state of +2, making the zinc atom divalent in all compounds. Figure 1.7 depicts a typical zinc atom with the two electrons in the outer orbit shell. For zinc, there is always two covalent bonds formed when the zinc atom combines with other elements.

Oxygen readily joins with zinc to form ZnO and $Zn(OH)_2$, with oxygen alone making a double bond and the two hydroxide combinations each with a single bond.

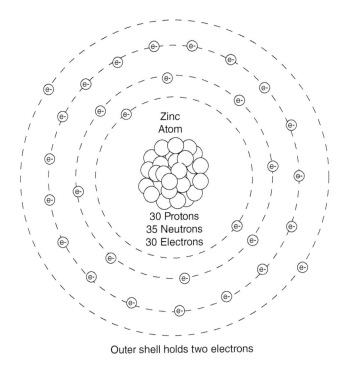

FIGURE 1.7 Zinc atom with two electrons in the other shell.

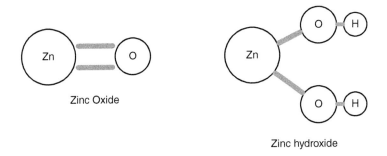

High-purity zinc is a strong oxidizer and when exposed to the atmosphere quickly tarnishes and forms the oxide and hydroxide. The standard potential of zinc can be expressed thermodynamically[4] as:

$$Zn \leftrightarrow Zn^{2+} + 2e^- \quad E0 = -0.763 \text{ v}$$

This represents a strong drive to combine with other elements.

---

[4]Zhang, Z.G. Corrosion and Electrochemistry of Zinc. Plenum Press, NY, 1996.

HISTORY

The discovery of zinc as a metal is credited in the West to the Swiss alchemist, Paracelsus. Dr. Paracelsus, as he was known because he was a physician and a philosopher as well, in 1526, described a metal he called *zinek,* as one of the seven known metals. Paracelsus lived around Basel and wrote extensively on various subjects. He is credited, among other things, as the father of toxicology.

Zinc ore was mined in Germany for the making of brass in the region around the Harz Mountains. The nearby town of Goslar, Germany, was a center of mining and zinc mining existed from around 1550. By 1650, a large-scale zinc ore production and refinement was underway. The mines around this region produced iron, silver, copper, lead, and zinc.

The process of refining the metal still was a mystery to the west. China and India would supply the metal in a refined state to European companies for producing brass by alloying with copper. Eventually, by the middle part of the eighteenth century, zinc-mining operations in Sweden and the region around Silesia would become important sources for the ore.

In the early 1700s, Bristol, England, the Bristol Brass Company would import zinc from India. William Champion, the son of the founder of the company, created a method of smelting his own zinc using a process notably similar to one developed centuries earlier in India. The company previously would import the zinc to make its brass plates, now it could produce and refine its own zinc from ore. William Champion has been credited with the early manufacture of industrial quantities of zinc.

Champion saw how the metal workers in India were extracting zinc from the ore pyrometallurgically by adding a distillation process to capture the fumes and condense the zinc oxide. The zinc was heated to turn it into vapor, the vapor would condense on the cooler walls of a chamber similar to the way it would condense on the cooler stone walls of the alchemists flue. This condensed zinc was zinc oxide. The key was to remove the oxygen by adding charcoal to the heated chamber and this would remove the oxygen from the zinc oxide creating carbon dioxide and leaving the zinc as a lump of metal.

In those days, brass was the main product that set the demand for zinc. Brass was used to clad the hulls of English sailing ships. Muntz metal, an alloy of copper and zinc, contains 40% zinc. Developed specifically as a cladding for ship hulls in 1832, Muntz, named after its inventor, George Fredrick Muntz of Birmingham, England, replaced copper as an anti-fouling cladding on the hulls of oceangoing ships. Because it had zinc, it was significantly cheaper than pure copper and still would protect the wood hulls from teredo shipworms. Muntz metal was also much stronger than copper, and the zinc lowered the melting point to 904°C from 1085°C for copper.

Zinc is a metal that has been in and out of art and architecture over the years. Since its discovery, or more so, since the time when casting and rolling into sheets, zinc has found use in architecture. The skyline of Paris is a testimony to the beauty and durability of the metal. Napoleon, around 1805 instructed the chemist Jean-Jacques Dony to develop the rich mines of zinc ore in the Vielle-Montagne. Dony developed a method of refining the ore using a horizontal distilling process that involved a series of retorts set into a furnace. The ore would be roasted, and the zinc fumes would be released and condensate, forming molten zinc. Figure 1.8 is a diagrammatic representation of the retort process.

12   Chapter 1   Introduction to Zinc

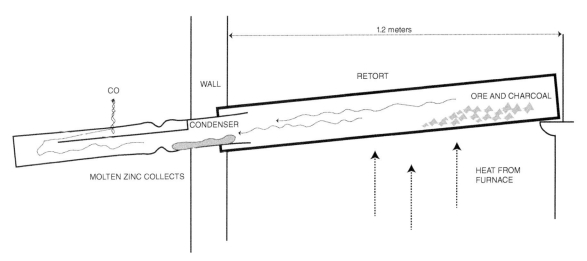

FIGURE 1.8   Diagram of a Horizontal Retort.
Source: Developed by Dony.

Dony set these retorts in series within a furnace and required a vast amount of heat energy. The process remained in use until the first half of the twentieth century. First, the reaction required the temperature within the retort to reach 1100°C or greater for the chemical reaction to occur between ZnO and carbon. The carbon was introduced from the charcoal, which would burn, creating carbon dioxide gas. Further heat would be applied to vaporize the zinc. Zinc oxide would form as a vapor, and when it combined with the heated carbon, the oxygen would be stripped away and vent out as carbon dioxide. The zinc would be condensing on the cooler portion of the retort and collect along the bottom as liquid metal.

These retorts would be set into arrays and charged with ore and charcoal.

With this new source of the metal, rolling into sheets and plates was possible using rolling techniques perfected with copper and iron plates. The first zinc-rolling mill for sheets of zinc was developed by Dony in 1812, making this metal available as an architectural cladding material to compete with copper and tin-plated steel.

Zinc has a long history in art and architecture. Its use as an alloy metal with copper to make brass was well understood by the Romans and Egyptians, who were attracted by the allure of the golden color the addition of the mineral calamine made with copper. At least they understood that something in the ore would interact with copper and produce brass. Zinc as a metal was unknown to early civilizations because it could not be separated from its ore as other metals. When copper, lead, tin, or iron ore were heated the metal would fall to the bottom of the furnace but zinc would boil and turn to vapor.

In China and in India, early metalworkers found ways of isolating zinc by roasting the ores in crucibles with charcoal and then allowing them to cool. Zinc would separate in small lumps where it could be collected and remelted.

History 13

FIGURE 1.9 An example of modern Bidriware made from zinc.

In the fourteenth century, there was an artform called Bidri that used hammered copper and zinc forms with incredible inlay artwork. Figure 1.9 is a modern example of the artwork. Bidri is a product that originated in south central India and is attributed to the Bahamani sultans in the fort city of Bidar. They used the process of engraving and repoussé to produce elaborate designs in metal bowls and plates as well as the bases of hookahs.

They would often inlay other metals such as silver or gold and then darken the background metal with sulfide compounds and polish off the top sections.

This area of India is still a major center for manufacturing unique metal work. Today, brass, copper, and zinc are still handcrafted here in the old tradition. It is important to note, these incredible art pieces were created from zinc–copper alloys, where the zinc was 4 times the amount of copper in the base metal to as much as 16 times. Zinc was being produced in large quantities in India as early as the fourteenth century.

There was no known process of producing zinc in Europe until several centuries later. In 1982, an archeological study of the mines in the region around Zawar in Rajasthan was undertaken by a British-Indian research group. They found intact furnaces and clay retorts that indicated smelting of zinc on a significant scale had been underway centuries ago[5]. The clay retorts were positioned at an angle. The neck-down area was lower than the enlarged section and positioned through a wall of clay and stone into a cooler chamber – much the same way as William Champion and Jean-Jacques Dony arranged their retorts to capture zinc vapor and condense it to create a pure form of zinc. William Champion traveled often to India, and apparently he studied the Indian process and brought it back to England.

---

[5] Jagdish Mittal, *Bidri Ware and Damascene Work in Jagdish and Kamla Mittal Museum of Indian Art*, Hyderabad, 2011, p. 39.

Other examples of zinc used in the far Eastern cultures predate the arrival of the metal to Europe. Zinc was rarer than copper and iron in these early times, and the utility of the metal was not yet understood until larger quantities could be produced.

In Europe, once more intense refinement of the metal took form and quantities of the metal became available, artists and artisan began to understand certain beneficial characteristics. One was the low melting temperature, much lower than copper or iron. Once melted, it had good fluidity and could be poured into simpler molds and achieve good detail.

Zinc-cast statues date back to the mid- to late-1700s in Europe, where it was extensively promoted for use in northern Europe. In Prussia, it was used on buildings and ornament for the new capital city of Berlin. Karl Friedrich Schinkel, the architect, artist, and city planner pushed for its use in statuary and building ornamentation in the early 1800s, where the silvery blue metal was used to adorn the new Prussian capital.

In Paris, one has to marvel at the silvery roofs and ornamentation that distinguishes that city. The Baron Hausmann, Prefect of the Seine Department of France under Napoleon III, undertook a vast redevelopment of the famous city in the mid-1800s. This started the cladding of the famous mansards of Paris. Supposedly, the Baron Hausmann had a relative in the zinc-mining business. It could also have been that Hausmann wanted to have a crème colored stone used for the walls of his building design and the use of copper may have led to staining. One of the great benefits of zinc is that its oxides do not stain adjacent materials.

One of the main sources of zinc was the mine, La Vielle Montagne in Kelmis, called *La Calamine* in French. This area, on the border of Germany was the source for much of the zinc used France at the time of this adornment and reconstruction of Paris. La Vielle Montagne started in the 1400s as a source for zinc used in manufacturing brass. The La Vielle Montagne Zin Mining Company was formed to supply Paris with the zinc needed to redevelop the city under Hausmann. The company became VM Zinc and is one of the largest suppliers of zinc in the world.

The zinc mines around Vielle-Montagne, had been in use since Roman times and this readily available ore was ideal for making a statement for France. As early as 1815, some of the first roofs of Paris were being clad in this silvery metal, zinc, and today close to 90% of the roofs of the great city are still covered in zinc. UNESCO, the United Nations Education, Scientific and Cultural Organization is considering making the zinc roofs of Paris a World Heritage.

The Prussian source of zinc was the area known as Silesia. Silesia, a region in present-day Poland, produced zinc that was known for its low sulfur. Very extensive manufacturing of zinc products took place in this region. The central part of Europe mined and produced much of the zinc used in the world during the 1800s.

It was soon discovered that coating iron in molten zinc would provide galvanic protection to the iron and later steel. By 1830, coating iron with zinc was in wide use throughout Europe. Later that century, steel was invented and overtook iron as a building material. As the less corrosion resistant steel came into major use the later part of the century, coating steel with molten zinc as a sacrificial layer became a major enterprise that continues today. The vast majority of zinc used today is to protect steel by hot dipping in baths of molten zinc.

The largest industrial companies producing zinc for architectural and industrial use were centered on this central region of Europe. Europe is the major producer of rolled zinc even today and some consider zinc the "European metal." In the late 1700s major zinc works in Bristol, Sweden, the upper Silesia region of Prussia and near Delach in Austria were in full operation, for the most part involved with alloying with copper to form brass.

After Dony's work in developing smelting and sheet production processes in Liege, Belgium, the Vieille Montagne Zinc Company was established in 1851 and became one of the largest producers of zinc sheet and ingot. Around the same time in Germany, along the Rhine in Westphalia zinc production also increased and formed one of the largest zinc production companies still in force today, Rhinezink.

The first use of rolled-zinc sheet occurred in Liege, Belgium, where in 1811 the roof of the Church of St. Barthelemy was covered with thin rolled plates. Also, in 1812, the tower steeple of the restored Cathedral of St Paul in Liege, was roofed with these first rolled-zinc sheets.

In 1837, the famous St Charles Hotel in New Orleans had a dome and roof clad in zinc sheets made from imported zinc sheeting from Vielle-Montagne to complement the white marble of this Grecian-style palace. This was the first use of architectural zinc in North America.

In North America, there was a period of time when the metal fell out of favor as a surfacing material in architecture and as a metal for sculpture. Knowledge of the metal waned. From around 1930 to 1990, the expertise in the use of zinc was deficient in North America, while in Europe it was taking a stronger position as a metal of architecture due to production practices that developed in France and Germany after the Second World War.

Early in North America, zinc was considered a waste byproduct in the production of lead. Lead and zinc were often mined together. It wasn't until the mid- to latter part of the 1800s that the metal zinc became valuable as a coating for steel and as an alloying element to add to copper to make brass bullet cartridges. Lead, of course, was the material for bullets. Around 1872, zinc became a commodity of interest as steel production took off in the United States after the civil war.

The need for galvanizing became critical with the arrival of steel. Steel was now being produced inexpensively as the Bessemer process advanced in development. Engineers were designing structures with steel and galvanizing, or coating with zinc. It was long known that zinc is as an excellent corrosion inhibitor for steels. The patent for galvanizing steel was filed in 1837 by the French Chemist Stanislas Sorel but the notion of zinc as a sacrificial material was described by Luigi Galvani and Alessandro Volta in the late eighteenth century.

In any event, once the economy of steel production was established, the need for zinc as the protective coating expanded and mining of zinc in the Americas began in earnest. Zinc mining and smelting expanded in various areas around the United States. Zinc deposits in Wisconsin, Illinois, and Missouri were exploited as the smelting process improved. In the late 1800s and early 1900s, American zinc-mining dominated supply of ore for the world. The large European companies had exhausted much of their supply of high-quality ore and turned their attention to importing zinc from American mining interests. The area around Joplin, Missouri, into Kansas and down to North Eastern Oklahoma became the center for zinc mining due to high-quality *jack,* as it was called. The Joplin, Missouri, area became the main supplier to the world as several million tons of high-quality ore was mined in this region. Every major rail line in North America passed through this region to transport zinc and lead ore eastward. Poor miners became wealthy with the rich finds of this metal ore they called jack.

A process of using vertical retort condensing was invented by the New Jersey Zinc Company. This process, was a modification of the horizontal retort developed by Dony in Belgium and was in wide use until the 1960s. In 1858, two immigrants, Fredrick W. Mattheiessen and Edward C. Hegeler, opened a zinc smelting plant in LaSalle, Illinois. Energy needed for the smelting operation was readily available and transportation on canals and rivers opened this region to the world. Before long the La Salle Zinc Works was producing 4,000 tons of zinc a year, making this one of the largest producers of refined zinc in the world. The La Salle Zinc Works continued to expand and supply zinc to galvanizing facilities for the burgeoning steel operations in the United States. They opened a rolling mill to produce slabs of zinc and eventually sheet and strip zinc. From an architectural context, they produced a sheet and strip zinc for roofing and flashing along the lines of what was being produced in Europe in the hopes of expanding the market for zinc in the United States in a similar fashion to its popularity in Europe. The product they produced was called *Titanaloy* because it was an alloy of zinc with small amounts of titanium added along with copper. It was introduced to the architectural market in the 1960s. The alloy gave zinc sheet improved creep resistance and improved strength. The alloy had been developed and patented by the New Jersey Zinc Company in 1949 and was licensed out for use by others 10 years later.

The Ball Metal and Chemical, a division of the Ball Corporation famous for their Ball Mason Jars, was another producer of zinc sheet in the United States. The Ball Corporation wanted corrosion resistant lids for their glass jars, and zinc, with its malleability and decent corrosion resistance, would suffice.

Their product for the architectural industry was called, Microzinc. Introduced in the early 1970s, this preweathered zinc sheet was alloyed with copper and titanium, similar to LaSalle's Titanaloy. The Ball Corporation spun off several of its business entities in the early 1990s and one was the zinc

TABLE 1.1 Common Minerals of Zinc

Mineral Name	Chemistry
Sphalerite (also called Calamine)	ZnS
Marmatite (variety of sphalerite)	$(Zn,Fe)S$
Hemimorphite	$Zn_4Si_2O_7(OH)_2 \cdot H_2O$
Hydrozincite	$Zn_5(CO_3)_2(OH)_6$
Smithsonite	$ZnCO_3$
Willemite	Zn_2SiO_4
Zincite	ZnO

production and rolling line in Tennessee. This became the Alltrista Zinc Products Company then later the Jarden Corporation in 2001. Today it operates as ARTAZN and is one of the largest producers of zinc strip and zinc-based products in the world. Still operating out of Tennessee, ARTAZN produces sheet and coil for the architectural market as well as manufactures the blanks for coinage used by countries around the world, including the US penny. The penny is zinc with a thin jacket of copper plate.

ZINC MINERAL FORMS

Zinc ores, like that of other heavy metals, are mostly below ground. The common zinc ore is sphalerite, sometimes called zinc blende in the United States and calamine in Europe. Sphalerite is composed of approximately 5% to 15% zinc. Table 1.1 shows the common mined minerals of zinc and Figure 1.10 shows several of the common minerals. The reddish center mineral is zincite, the top left and right minerals are smithsonite, while the black crystalline mineral at the bottom is smithsonite, "jack," from Joplin, Missouri.

Zinc is not as abundant as other elements. It comes in around 24th, just ahead of copper as elements that make up the Earth. Zinc is always found with other elements such as lead, copper, and sulfur. As the Earth formed, zinc combined with sulfur. This mineral was heavy compared to other substances and sank below the surface. Large deposits are mined in the Americas, Australia, and Asia where the ore has risen from tectonic activity over the millennial. Approximately 13 million metric tons of zinc ore are mined each year. Table 1.2 shows the comparison with other metals mined around the world.

Because zinc ores are relatively low in zinc, they have to be concentrated by crushing the rock and then floatation to remove other materials. Concentrating the ore increases the yield to around 60% zinc. Other metals, such as lead and copper, are removed as well. The concentrated ore is sent to the smelter where one of two methods are used to refine the zinc. The pyrometallurgic process, which uses the system of roasting and condensing or electrolytic refinement, which also begins with roasting but uses electrolytic processes to remove the zinc. Nearly all the zinc produced today involves the electrolytic method.

The electrolytic method also known as the Roast-Leach-Electrowin (RLE) process extracts the zinc by electrolysis. Essentially zinc oxide is produced from the concentrated zinc sulfide by roasting the ore at high temperature. This creates zinc oxide and sulfur dioxide. The sulfur dioxide is converted to sulfuric acid, which is then used in the leaching and electrolysis step. The zinc oxide is dissolved in the sulfuric acid in several concentrated steps. Other impurities are removed by a process of cementation. The concentration must be as pure as practical in order to efficiently pull the zinc out of solution. The impurities, often other metals, are also collected.

Once the concentrate is sufficiently pure, electrolysis begins. Here the zinc is removed from solution by electrolysis in the sulfuric acid electrolyte and deposits on the cathode. Oxygen forms at the anode. The zinc is stripped from the cathode, melted, and cast into ingots.

FIGURE 1.10 Various ores of zinc.

ZINC IN ART

Zinc as an artistic material or architecturally aesthetic surface was slow to be established in the United States. In the late 1800s there was a movement where statues and monuments were manufactured from zinc. The movement began in Europe, mainly France, where it was called *la statuomanie*, or *statue mania*. Statues proliferated throughout Europe and this eventually came to the United States as the Victorian era of architecture set in. The arrival of the movement coincided with the end of the American Civil War, and across the country, monuments to the solders of the war became commonplace. In addition to commemorating the civil war solder, state capitals, cemeteries and wealthy landowners adorned their buildings and gardens with sculpture. This, coupled with the

TABLE 1.2 Metal Mining around the World

Metal	Metric Tons Mined
Nickel	2.7 million
Zinc	13 million
Copper	20 million
Magnesium	29 million
Chromium	44 million
Aluminum	64 million
Iron	2.5 Billion

Source: Based on 2020 USGS Mineral Commodity Summary

Gothic and Italianate revival styles, ushered in with the Victorian era, opened zinc to the fashion of the time, with elaborate zinc castings or stamped zinc sheeting for both statue manufacture and fenestration on Victorian homes.

Figure 1.11 shows a typical Victorian era architecture that swept across the United States in the late 1800s. These homes were adorned with metal features made from tin and zinc, much of the work imported from Europe.

Copper alloys, such as gunmetal (85-5-5-5), with 85% copper, 5% zinc, 5% lead, and 5% tin, were used by sculptures during this time as well, but zinc played an important role. Zinc required less energy to melt, had a pleasing color that did not need to be patinated and protected with wax, and was affordable. As it weathered, it did not stain light colored stone or brick.

Zinc sculpture flourished during the late 1800s and early 1900s as patterns from European sculptures were taken and molds were generated from bronze castings to allow multiple zinc sculpture replicas to be made. Often, the zinc was treated to replicate the appearance of other materials such as bronze or stone. Plating copper to zinc was common and advertised as an alternative to bronze sculpture. Figure 1.12 shows a copper-plated zinc statue produced as a memorial to the union soldiers of the Civil War. There are many similar replicas of this same casting owing to the reuse of the molds.

A method of electroplating the zinc with copper came into widespread use as a way to obtain "bronze" for a more economical price. A true bronze sculpture might cost a hundred times that of a zinc sculpture. Sculptures made from zinc were quick and simpler to cast and the detail was similar.

On close inspection, these slush cast sculptures lack detail a similar bronze might have. Figure 1.13 is the face of a "Dough Boy" from World War I. This particular sculpture was slush cast numerous times from the same mold. The detail was good enough for the many replicas produced and shipped across the United States. As you can see, it too was plated in copper to mimic bronze.

There were numerous replicas of the Dough Boy manufactured by two competed firms at the time in the United States. You can find these copper-plated zinc sculptures in cemeteries and memorials all across the country.

Three methods of creating zinc sculptures were in use during the late nineteenth and early twentieth century. Sand casting, similar to cast methods used on copper alloys, was one method. A second process is known as slush casting. This method, similar to permanent mold type casting, involved pouring the molten zinc into the mold, allowing it to cool briefly so only the zinc at the interface of the mold wall cooled, then pouring the excess, still molten, metal out. This method was quick, and the outer surface would capture detail of the mold. Stiffness was achieved by thickening the walls. On occasion, several molds would be produced for a large assembly and then joined by

FIGURE 1.11 Vaile Mansion. Example of early Victorian Era homes in the US.
Source: Library of Congress

Zinc in Art 21

solder or by "welding" them together by pouring a band of zinc in segments along the joint between larger elements. Figure 1.14 shows the "Goddess of Liberty" statue erected on the top of the State Capital dome in Austin, Texas. This large sculpture was designed by the architect Elijah E. Meyers and cast in 1888. The sculpture is over 5 m tall and assembled from 80 unique cast panels. The panels were cast in zinc and assembled to an internal iron armature. It remained on the dome for nearly 100 years before damage from storms and fractures from thermal movement required it to be removed and replaced.

The third method involved stamping zinc sheet. The zinc sheet was heated and pressed into wood molds. Matching male and female dies were used to create the thin zinc panel. These would

FIGURE 1.12 Copper-plated zinc sculpture.

then be assembled by soldering the plates. Large sculpture could be created this way by incorporating an internal structure.

Initially, many of the sculptures were imported, but this quickly changed. Several enterprising companies set up foundries to create their own sculpture, often copying the designs of European artists.

The Monumental Bronze Company from Bridgeport, Connecticut, sold zinc statues in the late 1800s and early 1900s. They called the metal "white bronze," perhaps to give the statues a level of mystery and to ask a higher price.

FIGURE 1.13 World War I Dough Boy slush cast zinc. Copper plated.

Another company, M.J. Seelig and Company, established a foundry for casting zinc in Brooklyn in 1851. He is credited with casting the first zinc artwork in America for a Fair in New York in 1852.[6] The Seelig firm was prolific in casting zinc statues. Some were modeled after works from Europe, others were created by Moritz J. Seelig himself. The Seelig Company sold its sculptures through the J.W. Fiske Company and the J.L. Mott Iron Works Company. Each had catalogs of various sculptures for sale and marketed these across the United States. Some of these early statues had been copper plated and sold as "finished in bronze" in an attempt to add some panache. Others were

FIGURE 1.14 "Goddess of Liberty." Texas State Capital Building. Erected in 1888.
Source: Philip Arno Photography/Shutterstock

---

[6]Grissom, Carol A., Zinc Sculpture in America 1850–1950.

painted, like the "Goddess of Liberty," shown in Figure 1.14. This zinc sculpture was painted white to look like marble. Others were left in the traditional blue-gray color.

The supply of zinc for statues in the United States proliferated for nearly a century, perhaps peaking after the First World War. There were a number of foundries involved with producing sculpture from zinc in the United State as well as companies involved with the importation of sculptures from Europe. Many of these sculptures are still performing well even though they have been exposed for years.

Cracks from the condition known as creep are apparent in a few sculptures. Often, the fissures are mistaken for open seams due to freeze–thaw cycles. Creep is one of the major issues faced with unalloyed zinc. Creep is a condition where under sustained loading, even light loading such as the weight of the metal, deformation, and elongation occurs. Creep was an issue that had to be overcome if zinc was to be used as an architectural cladding. This was overcome in the 1960s as alloys were developed to give zinc better strength and the ability to resist creep.

ZINC AS AN ARCHITECTURAL METAL

Across Europe, zinc was considered a metal of architecture and ornamentation for buildings. Major cities, Paris, Berlin, Brussels, and others incorporated zinc in the architecture that defined the city. The Prussian architect Karl Friedrich Schinkel introduced zinc as an architectural metal in his designs for the city of Berlin. He was a city planner, artist, and architect who applied zinc in various ornamental forms on his designs in the early part of the nineteenth century.

France of course continues to adorn its buildings in Paris with zinc. From its elaborate mansard roofs to bay window dormers, cresting and other relief that grace the buildings of the city. Figure 1.15 shows a few of the beautiful zinc roofs that give France its character.

Still today, zinc is considered a major architectural metal for some of the most stunning architecture in Europe. Daniel Libskind chose zinc to clad the intricate shapes of the Jewish Museum in Berlin and the Filix Nussbaum Haus in Osnabrück, Germany.

The Winery Cantina de Il Bruciato, in Carducci, Italy, is an excellent example of the stunning use of zinc on modern architecture. See Figure 1.16. The design firm, Fiorenzo Valbonesi Cesena created an amazing addition to the Winery, using Rhinezink's preweathered zinc called Graphite Gray.

In the United States, the use of zinc was slow to develop. Initially, in the beginning of the twentieth century, zinc had a few inroads beyond statuary. Stamped panels, building ornamentation and other features, often painted were made from zinc.

Figure 1.17 shows the Folly Theater in Kansas City, constructed in 1900 and then renovated in 1980. When the metal work was taken down to be repaired, the workers were initially confused. The metal was different than the copper or terne-coated metal used on many of the buildings constructed in that era. Zinc had disappeared in the United States as a building material, and those who had worked with sheet metal for decades had no idea what this metal was. It wasn't magnetic, it could be soldered, it was dull bluish-gray, but it wasn't stainless steel. Zinc had fallen out of the vocabulary of sheet-metal workers for at least a generation in the United States. Replaced with lead-coated copper, terne-coated stainless steel, the blue-gray metal was for the most part completely unknown.

The upper cornice, capitals, railings, and ornamentation around the windows of the theater shown in Figure 1.17 were made from zinc and painted. The balusters were spun zinc and soldered together. Other elements were stamped and joined with soldered seams. For the most part, the metal was in good condition, a few open seams and missing elements, but the zinc had performed quite well even though the building had suffered decades of neglect.

There are various reasons, many psychographic in nature but mostly due to a lack of knowledge and understanding. Zinc as a building material was not especially promoted prior to the development of the zinc–copper–titanium alloy that entered the architectural picture in the early 1970s. Galvanizing was the main use of zinc and still is today, with almost 80% of the zinc produced going to protect steel either in automobiles, structures, or appliances.

FIGURE 1.15 Zinc roofs in Paris.

FIGURE 1.16 Winery Cantina de Il Bruciato designed by Fiorenzo Valbonesi Cesena. Image courtesy of the architect.

There were companies, like the Mesker Brothers of St. Louis, who touted the use of galvanized steel and galvanized iron castings for building facades. Figure 1.18 shows a bank façade created as a kit by the Mesker Brothers. Similar facades can be found in small towns across the United States where various parts could be assembled around the entryways and streetscapes.

Galvanized steel, both painted and unpainted, were the extent of zinc usage by architects for decades. The *Architectural Sheet Metal Manual* developed by the sheet metal industry in North America was silent on zinc. In 1929, the National Association of Sheet Metal Contractors published

FIGURE 1.17 Folly Theater in Kansas City, Missouri.

a beautifully detailed book titled, *Standard Practice in Sheet Metalwork*, no mention of zinc other than galvanized iron. The Sheet Metal Publication Company of New York, published, *The Universal Sheet Metal Pattern Cutter*, Volume II, discusses zinc in the same light as copper for assembling roofs, but makes no distinctions. On page 337 of the volume, zinc sheet roofing is listed as weighing 8 pounds per square foot (39 kg/m^2), as compared to 1 pound per square foot (4.8 kg/m^2) for copper. This would put zinc at a thickness of over 0.25 inches (7.5 mm). Granted, often early unalloyed zinc sheet was thick, much thicker than the copper–titanium alloys we use today. This was either a response for concerns of mechanical creep or a lack of knowledge of the metal. Most likely, the latter.

There is very little said in publications on architectural conservation that discuss the use of zinc on American historic buildings. The US Department of the Interior, 1992 publication, *Metals in America's Historic Buildings*, Part I, by Margot Gayle and David Look, AIA, mentions zinc and says, "Although the popularity and use of zinc roofing in America varied greatly, its use was never as widespread as tinplate."

The 1976 publication, *Architectural Sheet Metal*, by SMACNA[7], lists the metals copper, lead coated copper, stainless steel, aluminum, terne, terne-coated stainless steel, galvanized steel, and lead. No mention is made of zinc. The lack of knowledge of zinc as an architectural metal from those that worked with sheet metal architecture and the design community is a clear indication of why the metal was not used or considered for use.

There were producers of zinc sheet in the United States. Mattheiessen and Hegeler's LaSalle Zinc Works produced the product, Titanaloy, and Ball Metal and Chemical had their sheet product,

FIGURE 1.18 Mesker Brothers Kit Façade made of galvanized steel.

---

[7]SMACNA stands for the Sheet Metal Air Conditioning National Association. This is the association of those contractors that work on sheet metal applications across North America.

Microzinc 70. Both developed in the 1970s as roofing and flashing sheet metal but neither made a huge impact at the time with the design community or the metal fabricators. There was a lot of mis-information and no promoting group like what existed for aluminum, steel, and copper in the United States.

It wasn't until the late 1980s when the large European zinc manufacturing companies set their sights on the vast North American market and began a marketing and education campaign in earnest.

More than 90% of the architectural use of zinc that exists today was built after this period. Zinc use as a viable architectural metal in the United States has flourished and grown exponentially in the last three decades as education, knowledge and experience of this enigmatic blue-gray metal has expanded within the design community and the metal fabrication industry.

HEALTH AND HYGIENE

Zinc is a necessary element for human and animal life. Approximately 15 milligrams of zinc, is needed each day. Next to iron, zinc is one of the most important metals needed for proper development and, like iron, we are often deficient and need supplements of zinc.

Zinc is used by the body to create the enzymes needed for cell growth and aids in liver and kidney functioning as well as healing, fending off fatigue and even the common cold. We take zinc lozenges that contain zinc compounds to ward off or at least reduce the effect of common viruses and nasal congestion.

Both zinc and its oxide are not considered harmful to humans or animal life in light and moderate exposure. Humans use zinc oxide as a skin ointment to protect from the sun or as topical anti-inflammation cream to combat insect bites and poison ivy. It is found useful to combat and heal diaper rash on infants, burns and as a colorant in some cosmetics. The low toxicity of zinc enables it to be used as sunscreen to protect against harmful ultraviolet radiation. The maximum amount of zinc oxide in these various ointments, per accordance with the Food and Drug Administration of the United States, is approximately 25%.

Zinc in drinking water will change the taste when levels begin to exceed around 40 mg/liter. If the water is carbonated, the zinc ions go into solution much faster. Zinc is amphoteric, so it is attacked by both acids and bases. Milk, soda, and lemonade, for example, will attack the zinc more rapidly than distilled or tap water. Leaving these liquids in a zinc container can lead to dissolution of the zinc at levels that will alter the taste and increase toxicity.

Taking too much zinc into the body by drinking water, juice, food, or dietary supplements can cause adverse health effects such as vomiting, stomach cramps, and nausea. Extended intake of zinc will affect the kidneys and pancreas. Excessive zinc intake can block iron and copper absorption, leading to deficiencies in these important elements.

Not getting enough zinc in the diet can be just as bad, however. Young children require zinc to grow and develop properly. Too little zinc will also affect the immune system.

The zinc ion in solution can be harmful to animal and plant life. However, the zinc ion is a rapid oxider and combines quickly with oxygen and other substances. The occurrence of the zinc ion would be rare. When zinc is in the soil it will usually combine with other substances. That is one of the reasons we do not see zinc in pure form naturally. Too much zinc will interfere with plant growth as the free zinc ion is toxic to plants. This is what happens around smelting sites. Zinc is usually the least of the concerns as these sites have other more harmful substances such as lead and cadmium. Plants require zinc in very small amounts, or they will not grow. Zinc is considered a micronutrient for the health of plants.

When working with zinc, the most significant issue develops when inhalation of the zinc oxide fumes occurs. This can happen when welding or soldering zinc or welding galvanized steel. It can occur when cutting galvanized steel with plasma or other high temperature methods or working with casting zinc. Any high-temperature process that puts zinc oxide in the air can create the problem. Inhaling the fumes, even small concentrations can induce flu – like symptoms that lasts several days. Aches, chills, and fever set in. The first sign is a metallic taste in the mouth or throat irritation. This condition is as old as the metal itself and has several names, "brass foundry ague," "zinc fever," "brass shakes," and "metal fume fever." Whatever the name, it is not something to experience.

Similar to other metals, the early mining practices were ill-conceived. There are a number of zinc mining sites across the United States that were used for over a century. Often, smelting operations were located near these sites. Dust, air, water and slag waste accumulated in these sites. Zinc by itself, is not particularly toxic. Zinc ore though, is often accompanied by lead, cadmium and arsenic. Early mining was centered around lead and the zinc was considered a nuisance and quickly discarded. In the mining area of southwestern Missouri, the zinc minerals were called "jack," "black jack," or "rosin jack," depending on its color. Lead smelting was easier and the market for lead was significant. It was easier to work with at the time and bullets, made from lead, were in high demand.

The lead contamination along with other metals such as cadmium and arsenic are the major issues, still in existence today long after the mining and smelting operations ceased. These large mining sites, often located near towns and rivers, operated for decades. The accumulation of waste and contamination are considered toxic and are all in some form of remediation by the government. Deep pits, collapsing mine shafts, and the effect on water quality are conditions being faced by communities in and around these old mining operations. Most of these antiquated operations are no longer in business but the scar left by the operations is still there.

Today, the zinc industry in North America and Europe take the environment and environmental impact of their efforts seriously. Recycling of existing zinc currently stand at about 30% and is expected to increase as new systems of recovery bring the cost of recycling down. Recycling zinc from sheet, scrap zinc, and stamped zinc is relatively straightforward.

Figure 1.19 shows the recycled blanks from making the penny. These blocks will be remelted and turned into more coins or possibly, an architectural surface.

Recycling zinc that has been coating via the galvanizing process requires additional processes where the zinc is captured out of the vapor and dust from recycling steel. There are several processes that involve taking zinc out of the fumes generated from the melting and recycling of steel. As the steel is heated, the zinc melts and vaporizes. From the vapors the zinc is removed and repurposed.

FIGURE 1.19 "Un-money." Recycled blanks of zinc off-fall from penny manufacture.

Another method is to remove the zinc by leaching the metal with acid and electrowinning to recover the metal out of the acid bath.

Using zinc as a coating of steel in the process known as galvanizing is considered an environmental and sustainable coating process. There is energy consumed in the mining and smelting operation as well as the heating of the molten zinc bath. However, there are no VOCs (volatile organic compounds) released and the life cycle impact of hot-dipped galvanized is very low. The steel and the zinc can be recovered and recycled. The steel structure will last longer and have far less maintenance when properly done as compared to paint coatings. The initial cost of galvanizing is very low as well.

THE ENIGMATIC METAL

Zinc is possibly the least understood and enigmatic as metals go in an art and architectural context. Not unlike the hidden metal in the mineral form to our ancestors. More clarity of what zinc is about is coming into focus. As Figure 1.20 shows all the various names for the metal as we tried to figure out what its real value was.

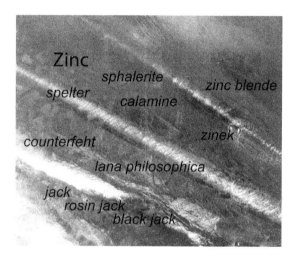

FIGURE 1.20 The names give to element 30 over the ages.

We often want to interpret the various metals through the same looking glass as far as how they are used. "If the metal can be cast, then it should have similar behavior," one might assume. "If the metal is produced in sheet, then it must form and work similar to other sheet metal forms," sheet metal is sheet metal to many.

Zinc, however, is different. It is both very corrosion resistant, yet it sacrifices itself rapidly to other metals if conditions are right. Zinc is easy to form, until the temperatures drop, then it can get brittle and crack when forming cold. Where most metals work harden from cold working, zinc relaxes. Zinc casts as easy as plastics, uses less energy than other metals, can even be injected into molds. Zinc can be elongated to extremes in superplastic forming. It sometimes acts more like a polymer than a metal.

The next chapters will explore this interesting metal and describe what distinguishes zinc from the other metals we use in art and architecture.

CHAPTER 2

Zinc Alloys

There has to be a better use for titanium than golf clubs.

Source: From Payne, Rob. Working class zero. Toronto, © 2003, Harper Collins Publishers.

INTRODUCTION

Early uses of zinc as cladding for roofs or castings were relatively pure. The retort system used to extract zinc from its ore produced a form of the metal that had good purity. There were traces of lead and cadmium that have low melting points as well and would be caught in the retort systems as they condensed along with the zinc. These elements are common occurrences in zinc ores. Today, in art and architectural alloys of zinc, the elements lead, and cadmium are kept at very small allowances within the alloy.

Various industries that use zinc have developed alloys that capitalize on different characteristics much like the industries that develop around the other metals. With zinc, though, there is not a large difference in properties between different alloys as there are with the alloys of other metals such as aluminum and copper alloys. Still, there are subtle differences that over the years were developed for specific industries. As an example, the architectural use of zinc today would be limited if not for the alloying elements, copper and titanium added to zinc. This has enabled zinc to perform mechanically in a similar fashion to other thin sheet metal used as surface cladding.

Another alloying element added to zinc is aluminum. Die-cast zinc is most always an alloy of zinc and aluminum. Aluminum is added to benefit strength while still enabling the metal to be cast. The alloys of zinc are designed specifically for an industry. The copper–titanium alloys are not used in casting, nor are the aluminum–zinc alloys used in roof cladding.

The marketplace for zinc is tighter than for other metals. The diversification of zinc into other areas is limited by the mechanical behavior of the metal. Zinc is weaker than other metals and this limits the metal to industries and systems not subject to high stress conditions. You would never think of using beams and columns made out of zinc, nor would an airplane manufactured from zinc operate very efficiently.

Zinc does not have an appreciable number of metals that can go into solid solution with it. Zinc can only except small amounts of other metals into its compact hexagonal structure. Gold and silver can form solid solutions with zinc up to around 10%, whereas copper, which accepts as much as 39% zinc to form brass alloys, can only be absorbed into the zinc matrix to around 3%.

Zinc forms several binary alloys. Zinc and aluminum is an important die cast alloy. Zinc and iron are critical for the interface of zinc to steel in the galvanizing process. Zinc and copper are binary alloys used to make the metal harder and provide the base alloy that is plated with copper to make the copper penny.

The major zinc ternary alloy is composed of zinc, copper, and titanium. The development of this alloy has greatly improved the thin-rolled-zinc product for architectural cladding.

Alloying elements are important but have limited effect on the mechanical performance of zinc. Improvement in hardness and creep resistance are the most important aspects. Cold working and heat strengthening have little effect on zinc due to the way the atoms fall back in alignment after plastic deformation. It essentially recrystallizes at room temperature once the deformation occurs. The exception here is the zinc, copper, and titanium alloy, which needs post rolling treatments to reduce some of the anisotropic behavior of the rolled sheet. Unlike other metals, though, zinc does not have the ability to establish temper in levels that create any appreciable changes.

The zinc alloys over the past century have undergone development and specialization much the same as other metals. Conventional foundries in the United States were slow to using zinc as a casting metal, even with the low melting point. The metal was not well understood, and cast alloys were considered more for the die casting of smaller custom parts. Slush casting was used by several specialized foundries in the late 1800s and early 1900s, but this fell away from interest after World War I in regard to art and architecture.

The low melting temperature and fluidity of the metal makes it an exceptional subject for high-detail casting. The art world in Europe saw this, and in the late 1800s to early 1900s used zinc to create some magnificently detailed work. Figure 2.1 is an example of highly detailed zinc casting.

ALLOYING DESCRIPTIONS

Zinc alloys are often known and described by various industry convention or the percentage of alloying constituents they contain. For example, the rolled-zinc alloy that contains copper and titanium is often described as Zn-0.8Cu-0.12Ti for an alloy with 0.8% copper and 0.12% titanium. This is similar for the cast alloys. For the cast alloys, such as ZA 8, this describes an alloy with approximately 8% aluminum. The cast alloys can be described in industry by AG40A or AG40B, which is the ASTM International designation, or by such trade names as Zamac, Zamak, Mazak, kirksite, or the initials ZA for zinc aluminum. Zinc, like all the other metals, has its own peculiar vocabulary.

FIGURE 2.1   Zinc casting of a boy and his dog.
Source: Courtesy of Zahner Metal Conservation

Over the years, the various zinc manufacturers sought to brand their specific alloys with names or numbers that distinguished their product from that of a competitor – Titanaloy, Microzinc 80, alloy 190, and so on. Today, the distinction is made by the color tone offered. Roano zinc™, Anthra zinc™, Glacier Gray™, Slate™, and other names that have some reflection or divergence from the metal itself. Thus is the way of marketing. In the end, it is zinc alloyed with copper and titanium with a branded surface finish name.

The Unified Numbering System (UNS) is used in this and the other books of this metal series. The UNS numbering system presents a logic to the numbering process. This designation begins with the letter Z for zinc and is followed by five numerals. The first numeral indicates the major alloying constituent and the balance of the number designations are for the different alloys within each category.

Table 2.1 shows the general categories.

TABLE 2.1 Zinc Categories in the Unified Numbering System

Z1xxxx	The number 1 stands for unalloyed zinc with various impurity limits
Z2xxxx	Zinc alloys containing lead as the predominant element
Z3xxxx	Zinc alloys containing aluminum as the predominant element
Z4xxxx	Zinc alloys containing copper as the predominant element

INGOT ALLOYS

The majority of the zinc produced each year goes to the protection of steel. Zinc and steel are particularly acquainted. There are no other metal that pair so well together. Galvanizing of steel for the building, appliance, and transportation industries use a significant quantity of the metal, and there are alloys designed for this specific application.

These alloys are provided in slab and ingot forms of the zinc. The alloys are defined by their purity and come in the grades listed in Table 2.2.

Each of these grades is delivered in solid block form to the hot-dipped galvanizing plant, zinc-casting facility, or zinc continuous cast facility. By starting with high-purity grades, better control of the final alloy is assured. These solid forms are the ingots of zinc alloy that are added to the galvanizing kettle to be melted and used to hot dip galvanize ferrous parts.

Ingots come in various sizes, from small 2 kg pigs (see Figure 2.2), to 25 kg (55 lb) blocks and small zinc pellets, depending on what is needed to replenish the galvanizing kettle. The alloy grades known as High Grade (HG) or Special High Grade (SHG) are used when high purity is required. Continuous sheet galvanizing lines and electro-galvanizing lines, brass production, zinc die-cast operations, and continuous casting lines require the high-purity grades.

The lower purity, Prime Western, is used in hot-dipped galvanizing when the parts are large fabricated assemblies. Typically, the purity is not as critical for these forms. Prime Western Grade (PWG) also comes in recycled grade and is labeled, PWG-R. There is an allowable lead and cadmium content in the PWG as well as other impurities. The lead is beneficial for obtaining adequate coverage of the formed parts when they are immersed in the molten zinc. The allowable lead according to North American standards is a maximum 1%.

TABLE 2.2 List of Slab Form or Ingot Grade Zinc Alloys

Alloy No.	Name	% Purity
Z13001	Special High Grade	99.99
Z15001	High Grade	99.97
Z19001	Prime Western Grade	98.65
Z18004	Prime Western Recycled	98.50

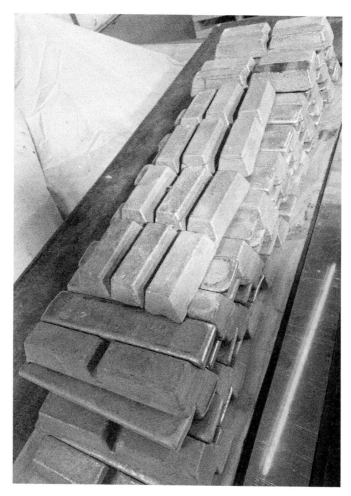

FIGURE 2.2   Small ingots of HG zinc alloy Z15001.

Protecting steel and iron parts by the sacrificial character created by the hot-dipped galvanized process occupies close to 50% of the zinc used worldwide.

Figure 2.3 shows the approximate overall percentage of zinc used in various industries.

Second to galvanized steel, zinc is used to make brass and other copper alloys. Using zinc to make copper harder and stronger has been underway for centuries, even before the metal was known to exist. As zinc is added, up to approximately 40%, copper alloys turn more yellow or golden and increases in strength and hardness.[1]

---

[1] More information on alloys and alloying effects on copper can be found in the book: L.W. Zahner, *Copper, Brass, and Bronze Surfaces: A Guide to Alloys, Finishes, Fabrication, and Maintenance in Architecture and Art (Architectural Metal Series)*. (Hoboken, NJ: John Wiley & Sons, 2020).

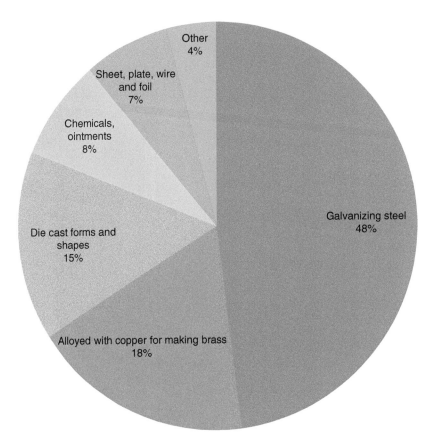

FIGURE 2.3   Uses of zinc.

Die-cast forms and shapes from zinc alloys are innumerable. They are used in the automotive, toy, computer, electrical, and any number of industries that utilize production casting with high accuracy. Zinc can be quickly cast into inexpensive parts. Zinc is easy to machine, and cleanup is minimal. The nonsparking aspect of zinc make it ideal for small gasoline engine parts as well as sensitive appliance parts. Figure 2.4 shows cast zinc small engine parts. Due to zinc's low-melting point die casting zinc can be a rapid and cost-effective method to produce nonmagnetic, corrosion-resistant near net shapes.

Zinc is corrosion resistant to most fluids and in most atmospheres. Acidic substances and strong bases should be avoided, but most other near pH neutral fluids and solvents will not harm zinc unless they remain on the zinc surface in a confined space where air movement is restricted. It is well-known that gasoline with ethanol will corrode zinc carburetors if left for several months. Ethanol absorbs moisture, and this moisture in a confined space will corrode the zinc surface on the interior of the carburetor.

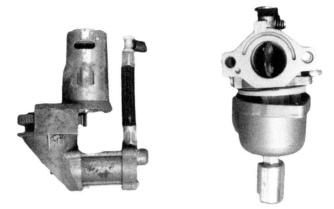

FIGURE 2.4 Examples of die-cast zinc used in small engine parts.

Chemicals and ointments use zinc in the powder form or combined with other substances. Zinc-rich paint, another method used to provide a protective coating for steels, uses zinc powders dispersed throughout a resin. Zinc white paint is created from zinc oxide. Additionally, zinc oxide is used to protect the skin against harmful ultraviolet radiation, treat acne, athlete's foot, and diaper rash. Zinc oxide is water insoluble and has a high refractive index, which makes it a great ultraviolet radiation (UV) absorber. Zinc glucomate is used as a throat spray, and zinc acetate is the chemical in throat lozenges that we take to fend off colds.

ZINC ALLOYS – ROLLED FORMS

Rolled products include plate, sheet, strip, and foil. The term *rolled* used to describe these flat forms is a reference used by the industry to distinguish this form of the metal from castings. This is the set of alloys where the primary architectural alloys reside. Rolled-zinc alloys are also the initial form used to make coinage and the common household battery. Many coins used around the world, in particular the US "copper" penny, are actually plated zinc. The zinc blank is punched to a designed thickness from zinc strip before being plated with copper in multiple, large rotating plating baskets. Millions of coin blanks are produced in a single day this way.

The common dry-cell battery makes up a large market for zinc. The batteries we use in our everyday devices are composed of an outer shroud of painted zinc. The inner surface and the exposed end make up the negative electrode of the assembled battery. Inside this shell is the electrolyte, usually a paste-like substance composed of manganese oxide and ammonium chloride. Within the electrolyte is the positive anode, usually composed of graphite and connected to the end cap. The end cap is separated from the zinc housing. The casings for the battery cans are cut from rolled strip into blanks called zinc *calots,* then impact extruded to form the can shape. Batteries make up a significant usage of zinc strip. Figure 2.5 shows the workings of a battery.

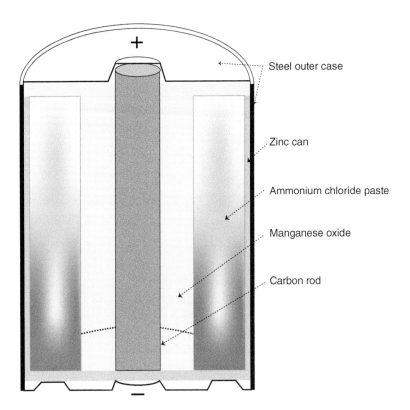

FIGURE 2.5   Diagram of a battery.

Zinc roofing, flashing, gutters, and zinc wall panels are made from rolled-zinc sheet. These are composed of the alloys specially designed for the rigors of environmental exposures. Prior to the 1960s, zinc roofing was made from unalloyed sheet. The zinc had trace elements from the refining and casting process. Lead, cadmium, and iron were common impurities found in early zinc sheet.

This early zinc sheet was subject to *creep,* a condition where the metal had a tendency to elongate under load. Creep is the name given to a time-dependent stain condition that occurs when the metal is under constant loading. Unalloyed zinc is weak. No amount of cold working would improve the strength or hardness. Unalloyed zinc is brittle and will crack if worked in temperatures below room temperatures. Early zinc was produced in thicker sheets to accommodate the weakness and creep tendency.

In 1944, the New Jersey Zinc Company patented an alloy of zinc with titanium and manganese. This showed improvement, but it was the work in 1949, in Germany by Dr. Erich Pelzel of Stolberger Zinc AG that changed everything for the advancement of zinc in architecture. Dr. Pelzel is credited with arriving at alloying of zinc with minute amounts of copper and titanium. He had numerous patents covering his successes in the development of zinc alloys. He was issued a US patent in the early 1970s on an alloy of zinc with copper and titanium. This alloy significantly improved strength without sacrificing ductility. Creep-resistant properties were elevated. He arrived at an alloy to be used for zinc sheet and strip that had an optimum combinations of properties dealing with creep resistance, ductility, and tensile strength.

Europe embraced the zinc alloy and industries developed around roofing, guttering and flashing in the early 1960s. In the United States, the large zinc producers, the Ball Corporation and the Matthiessen and Hegeler Zinc Co., produced roofing products in the 1960s using this alloy and introduced them to the architectural sheet metal market with products branded with the names Microzinc and Titanaloy.

Today, these copper–titanium–zinc rolled sheets are in use around the world, as the production processes have improved the quality and metallurgy of the final zinc sheet.

The major manufacturers are creating sheet in continuous casting machines where the zinc is melted, copper and titanium are added, and the alloy is passed through a set of moving ceramic coated belts that are cooled by water. The zinc solidifies into a thick plate that continues on to be reduced in a series of rolls. Figure 2.6 shows zinc plate exiting a Hazelett Machine.

Sheet zinc used in architecture, sometimes referred to as *rolled zinc,* differs from the other architectural metals in the way it is brought to market. Unlike stainless steel, copper, and aluminum, sheet forms of zinc are sold as a branded product rather than a commodity.

Zinc sold in ingot form falls under an alloying specification established by industry and is considered a commodity, whereas once the zinc is further processed, the various manufacturers establish the rolled form of the sheet as a product brand in order to differentiate. Note also, most zinc sheet and coil is supplied to the architectural world as preweathered – that is, the surface has undergone a treatment to form a matte, low-reflective oxide on the surface. Preweathering is a further market differentiator. The zinc produced by one company will not match the appearance of the zinc produced by another due to variations in the process steps, nor necessarily the alloy itself. Several different zinc manufacturers' preweathered and patinated products are shown in Figure 2.7. Each is of similar thickness. All zinc sheet used in the architectural cladding market is made from the zinc–copper–titanium alloys. The alloying constituents may change somewhat from the specification but not significantly.

ZINC ALLOYS USED IN ARCHITECTURE

In keeping with the other books in the Architectural Metal Series, the Unified Numbering System will be used to identify the alloy where possible. There are variations in alloying constituents from European products and the North American products of similar type, and these alloys do not match up precisely with the UNS constituents. Some of this is due to the branding and differentiation. The cast alloys have corresponding standards relating to the alloy-numbering process. For the architectural rolled plate and sheet, there are variations by manufacturer. The alloys fit particular categories established by industry, but often these alloys are made to order for the manufacturing entity. It is advisable to check on the alloying constituents and the mechanical properties of the particular supplier.

FIGURE 2.6 Zinc plate being continuously cast.

Zinc Alloys Used in Architecture   43

FIGURE 2.7   Examples of sheet zinc made from different manufacturers.

TABLE 2.3   Alloys of Zinc

UNS Alloy Number	Use (other than ingot)
Z30500	Alloy used in slush casting
Z33520	Alloy used in die casting
Z33540	Alloy used for making tooling
Z34510	Alloy used in slush casting
Z40330	Rolling
Z41320	Rolling

The alloys that will be discussed relate to the architectural industry. These are listed in Table 2.3, along with the associated product form and type. The diecast and slush cast alloys are not necessarily architectural but they do have a connection to art and potential for future applications.

The most common impurity found in zinc is lead. Zinc is often found in nature with lead, and traces of lead often find their way into zinc. Zinc can be refined by additional thermal treatments to reduce lead levels. Today, most uses of zinc have very low lead levels.

Other metals that are purposely alloyed with zinc or are found as trace amounts are listed in Table 2.4. Some of these, cadmium and tin for example, are kept to trace levels. Iron and zinc have a special affinity. Iron in very small amounts is often picked up by interaction with the mold walls or steel kettle walls. Die casting, one of the common uses of zinc, involves alloys of zinc and

TABLE 2.4 Common Alloying Elements in Zinc

(Al)	**Aluminum** is added to zinc die cast alloys to increase strength and to reduce grain size. Aluminum improves cast ability and fluidity of the casting. Aluminum above 5%, which is the eutectic limit, makes the alloy brittle.
(Cu)	**Copper** increases hardness and strength. Small additions of copper will greatly improve tensile strength. Copper also helps combat inter crystalline corrosion in castings. Excessive copper, however, can lead to brittleness at room temperature.
(Ti)	**Titanium** induces fine grain development. This aids in creep resistance, particularly when coupled with small amounts of copper.
(Fe)	**Iron** is often a trace element picked up during the process of casting. Zinc has an affinity for iron and tends to capture small amount of iron.
(Pb)	**Lead** is a trace contaminant. Commonly present during the mining of zinc. Lead quantities greater than 0.005% can have an effect on casting quality.
(Cd)	**Cadmium** also is an element that often accompanies the zinc ore. Quantities of cadmium in excess of 0.004% can have a detrimental effect on a die casting. Cadmium increases stiffness and improves etching and engraving quality.
(Sn)	**Tin** is a trace element. Tin in excess of 0.003% can cause subsurface corrosion in the die casting. Tin can also cause embrittlement in rolled alloys.
(Mg)	**Magnesium** in small amounts improves hardness.

aluminum. Aluminum up to 4% is typical. There is an alloy of zinc and aluminum that contains 22% aluminum and exhibits several interesting characteristics. Initially called Prestal® or ZA22, this alloy has high strength, nearing strengths of mild steel, but it can be molded and shaped and exhibits super plasticity. By super plasticity, this alloy can be deformed as much as 1000%, exhibiting properties of plastics by deforming and elongating without breaking.

Zinc with small additions of copper and titanium has transformed the metal into a major architectural metal. These alloys make up nearly all the rolled-zinc sheet form used in the building industry. Zinc–copper–titanium alloys strengthen and harden the metal to sufficient levels that it can be used as cladding material with specific engineering and mechanical attributes. The addition of very small amounts of titanium improves the creep resistance of the metal, which has allowed thinner zinc sheets to be used in many architectural applications.

WROUGHT ZINC ALLOYS

Wrought zinc are those products that undergo subsequent processing to refine the grain structure by mechanical alterations such as thickness reduction by rolling under pressure or extruding to create rod or wire from cast billets. These forms include the rolled forms mentioned previously.

Wrought zinc forms are as follows:

Plate
Sheet
Strip
Foil
Wire
Extrusion

Rolled zinc is the name given to the forms of plate, sheet, strip, and foil. There is at least a dozen rolled-zinc alloys, counting commercial pure zinc. In the architectural and art market, there are a few that are promoted by various manufacturers for use as a surfacing material. They are all very close in alloying constituents, with the distinctions in surface finish and preweathering techniques used by the various manufacturers of the sheet material. Commercial pure zinc is not used due to its low tensile strength, softness, and tendency to creep.

There is a soft alloy used as a flashing material for roofing applications. This alloy has replaced lead as a malleable, hand-formable material used as valley flashing or other moldable flashing. Alloy Z15006 is sometimes used for these applications because of its low strength and softness. Besides pure zinc, there are alloys of zinc with copper, zinc with lead, cadmium, and iron. There are alloys of rolled zinc with aluminum and magnesium.

The alloy almost exclusive to art and architecture is the zinc alloyed with copper and titanium. Figure 2.8 is an example of a roof surface made of rolled preweathered zinc. The color is a matte, blue gray.

FIGURE 2.8   Zinc rooftop made of rolled zinc.

In architectural zinc sheet, there are two types in common use, qualified as Type 1 and Type 2. These are defined in ASTM B69-13. The major difference is the range of allowable copper in the alloy. Type 2 has a wider range of expected elongation. This would allow some shaping and drawing improvement, but only slightly.

From an architectural or art context, the Type 2 has a distinctive color and sheen as compared to Type 1. Type 2 has a green-gray color tone versus a blue-gray. This has to do with the manufacturing process used to produce the sheet and the preweathering treatment utilized by the manufacturer.

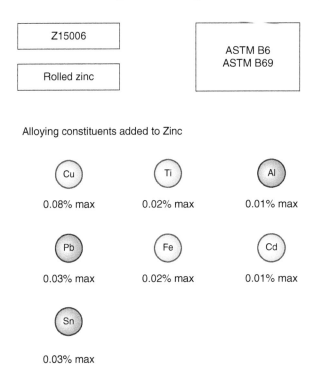

Alloy	Tensile Strength	Yield Strength	Elongation %(2in – 50mm)	Rockwell Hardness B
Z15006	14 ksi 97MPa	11 ksi 79 MPa	38	40

Alloy Z15006 is also known as commercially pure rolled zinc. It is 99.83% or higher in purity. The listed elements are allowable trace elements sometimes found in the makeup of this alloy. Z15006 is low in strength and soft. Not as soft as lead, still it can be easily shaped. This zinc has a very low tensile strength. Of the rolled-zinc alloys in common use, this has the lowest tensile strength. Alloy Z13004 is similar. This alloy has tighter purity levels and is as weak in strength.

There are other soft-rolled alloys of zinc used as flashing, separators, and battery casings. The HG rolled-zinc alloys are used in the battery casings to act as anodes.

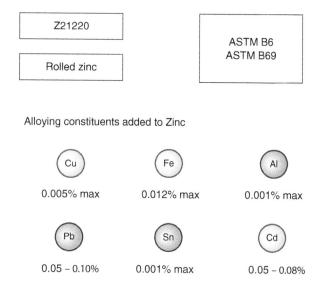

Alloy	Tensile Strength Range	Yield Strength Range	Elongation %(2in – 50mm)	Rockwell Hardness B
Z21220	150–170 MPa 22–24 ksi	124–138 MPa 18–20 ksi	30	–

Alloy Z21220 is used to make spun forms and deep stamp cylinders for batteries casings. This alloy is anisotropic and is prone to creep when subjected to loads.

This alloy contains lead and cadmium in very small amounts. It is not a common architectural alloy.

ARCHITECTURAL ROLLED ZINC

The architectural zinc marketplace is one of product differentiation. All the major sheet producers use a variety of the zinc–copper–titanium. The manufacturers of rolled-zinc sheet are found in many parts of the world. The oldest are in Europe. They have various brand names and usually qualify them as titanium bearing, like the panache product Titanaloy of old. This trade name, *Titanaloy*, was a product of the Matthhiessen and Hegeler Zinc Company of La Salle, Illinois. Once one of the largest zinc companies in America, Matthhiessen and Hegeler Zinc Company wanted to convey the strength and durability of this new alloy of zinc.

To capture the modernistic nature of design in the 1960s, why not emphasize the metal named for the Titans of Greek mythology? Prior to alloying with titanium and copper, zinc was considered soft and weak. It needed a new introduction.

Over the decades, many of these brands have come and gone as businesses and marketing efforts changed. A few of the trade names of rolled zinc with copper and titanium as two of the major alloying constituents are listed in **Appendix A**. Some of these have been absorbed or changed into other brands.

Saying the metal is made of titanium, even though the amount is quite small, gives an aura of toughness to the product. Titanium in trace amounts does help with strength, particularly creep resistance. Creep was a significant constraint to rolled zinc in architecture prior to alloying with titanium. When rolled out in sheets, the grains of zinc and titanium will align along the direction of rolling. This enhances the strength, yet still affords good ductility. Figure 2.9 shows a rolled-zinc panel that has been custom patinated.

Copper also gives the alloy added strength and improved hardness. Copper is in higher percentages than titanium in the rolled zinc used in art and architecture, but, I suppose, calling this new zinc alloy "copperaloy" would be a bit confusing and lacks the overtone afforded by the Greek deities.

FIGURE 2.9 Interior surface with custom Hunter™ patina zinc panels.

Architectural Rolled Zinc

TABLE 2.5 Rolled-Zinc Alloys

Alloy No.	Copper	Titanium
Z41110	0.08–0.20	0.70–0.12
Z41121	0.08–0.50	0.05–0.18
Z40301	0.50–1.00	0.70–0.12
Z41310	0.80–1.00	0.70–0.12
Z41320	0.50–1.50	0.12–0.50

From an engineering material context, it is important the rolled-zinc alloy of copper and titanium undergo prescribed mechanical and thermal treatments to achieve the creep resistance. These are posttreatments performed on the sheet at the factory, and these posttreatments refine the grain in the sheet.

The following alloys are discussed as a group (see Table 2.5). These alloys are anisotropic. Their mechanical strength characteristics vary considerably when considering with the grain and against the grain.

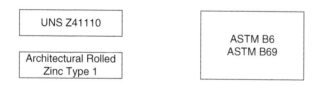

Alloying constituents added to Zinc

Cu	Ti	Al	Pb
0.08 – 0.20%	0.7 – 0.12%	0.001 – 0.015%	0%

Alloy	Tensile Strength Range	Yield Strength Range	Elongation %(2in – 50mm)	Rockwell Hardness B
Z41110	96–262 MPa 14–38 ksi	-	70–10	54–74

Chapter 2 Zinc Alloys

Z41121

Architectural Rolled Alloy 710

ASTM B6
ASTM B69

Alloying constituents added to Zinc

Cu	Ti	Al
0.08% – 0.50%	0.050 – 0.18%	0.01 max

Pb	Fe	Cd
0.01%	0.01%	0.005%

Sn
0.005%

Alloy	Tensile Strength Range	Yield Strength Range	Elongation %(2in – 50mm)	Rockwell Hardness B
Z41121	140 MPa 21 ksi	120 MPa 17 ksi	50	–

UNS Z40301

Rolled zinc

ASTM B6
ASTM B69

Alloying constituents added to Zinc

Cu	Ti	Al	Pb
0.5 – 1.00%	0.7 – 0.12%	0.001 – 0.015%	0.01 max

Cd	Fe	Sn
0.005 max%	0.01 max	0.03 max

Architectural Rolled Zinc

Alloy	Tensile Strength Range	Yield Strength Range	Elongation %(2in – 50mm)	Rockwell Hardness B
Z40301	190 MPa 27 ksi	150 MPa 22 ksi	50	–

UNS Z41310

Architectural Rolled Type 2

ASTM B6
ASTM B69

Alloying constituents added to Zinc

(Cu) 0.8 – 1.00% (Ti) 0.7 – 0.12% (Al) 0.001 – 0.015% (Pb) 0%

Alloy	Rolling Direction	Tensile Strength Range	Yield Strength	Elongation %(2 in–50 mm)	Rockwell Hardness B
Z41310	With grain	160–207 MPa 23–30 ksi	125 MPa 18 ksi	70–35	54–66
Z41310	Against grain	200–262 MPa 29–38 ksi	177 MPa 26 ksi	80–15	54–66

UNS Z41320

Architectural Rolled

ASTM B6
ASTM B69

Alloying constituents added to Zinc

(Cu) 0.50 – 1.50% (Ti) 0.12 – 0.50% (Al) 0.001 max (Pb) 0.010 max

Chapter 2 Zinc Alloys

Alloy	Tensile Strength Range	Yield Strength Range	Elongation %(2in – 50mm)	Rockwell Hardness B
Z41320	200–260 MPa 29–37 ksi	125–177 MPa 18–26 ksi	60–44	80

The properties of the rolled-zinc alloy containing copper and titanium vary only slightly from one to another. These alloys can have their mechanical properties influenced by posttreatment. To reduce creep, these alloys should undergo heat treatment after being cold worked. Compared to the other architectural metals, zinc alloys are not as strong from a mechanical context. Still, for most architectural and art cladding situations they perform quite well when designed correctly. Figure 2.10 shows the stadium in Barcelona, clad in rolled zinc.

Another factor with rolled-zinc sheet or plate is the variation of strength parallel with the grain or perpendicular to the grain. There is a markedly different strength profile when the metal is worked in the direction of the rolling process as opposed to across the sheet. "With the grain" refers to any loads being applied to the sheet or plate along the length, versus "against the grain" or across the sheet. Figure 2.11 shows when a load is applied to a rolled-zinc surface supported along the edges, the metal is weaker.

If the zinc surface is supported across the grain, the metal has better strength. This is called *anisotropy* – the metal is stronger mechanically in one direction versus another. Anisotropy is apparent in most rolled sheet metals. Zinc, because of its grain alignment, has more pronounced anisotropic bias and is inclined to be more rigid across the grain of the sheet (transverse) than parallel with the grain.

Essentially, if a panel is manufactured from rolled zinc, a design must consider how the zinc will react as the load is imparted to the formed zinc sheet. The zinc panel may perform perfectly well when analysis is done on the zinc section perpendicular to the grain, but parallel to the grain it may flex more because it is weaker.

FIGURE 2.10 Palau Sant Jordi, Barcelona. Designed by Arata Isozaki Marco Rubino/Shutterstock.

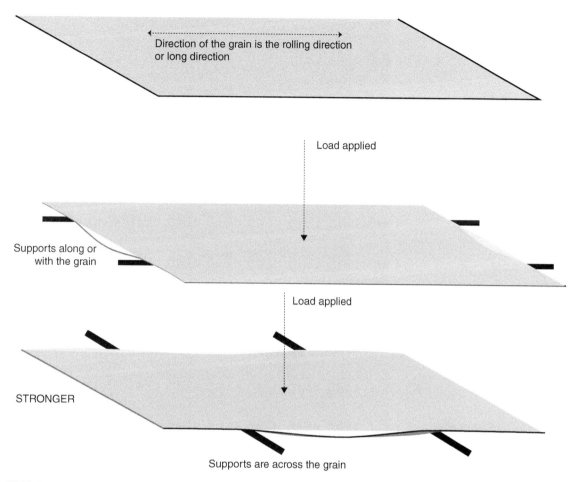

FIGURE 2.11 Anisotropic nature of rolled-zinc sheet.

FORGED AND EXTRUDED ZINC ALLOYS

Zinc alloys that are used in forging processes, surface coining processes, and extrusion are fine-grained alloys containing copper and titanium. Alloy Z35841 (ZA27) is another alloy that has good grain structure and contains 27% aluminum. This alloy is also described in the gravity cast zinc alloys. The common forging alloys and their alloying constituents are listed in Table 2.6. Mechanical properties of the forging alloys are listed in Table 2.7.

Korloy is a registered trademark of Cominco LTD. of Canada. The alloy is produced by several manufacturers to be used for forging and extruding. Alloy 190 is a trade name reference. This alloy is used to manufacture coinage. The US penny is currently made from this alloy and subsequently plated then forged in a coining press by the US mint.

Chapter 2 Zinc Alloys

TABLE 2.6 Alloying Constituents in Several Forging and Extrusion Alloys

Alloy	Common Alloy #	Al Aluminum	Mg Magnesium	Cu Copper	Fe Iron	Pb Lead	Cd Cadmium	Ti Titanium	Mn Manganese
	Alloy 190	–	–	0.7–0.9%	–	–	–	–	–
	Korloy 2573	14.5%	0.02%	0.75%	–	–	–	–	–
Z35636	Korloy 3130	–	–	1.0%	–	–	–	0.1%	–
Z35631	Korloy 3330	–	–	1.0%	–	–	–	0.1%	1.0%
Z35841	ZA27	25.0–28.0%	0.01–0.02%	2.0–2.5%	0.075%	0.006%	0.006%	–	–

TABLE 2.7 Approximate Mechanical Properties of the Forging Alloys

Alloy	Common Alloy #	Tensile Strength	Yield Strength	Elongation
	Alloy 190	152–200 MPa 22–29 ksi		35–70%
	Korloy 2573	345–517 MPa 50–75 ksi	310–448 MPa 45–65 ksi	20–14%
Z35636	Korloy 3130	241 Mpa 35 ksi	158 MPa 23 ksi	30%
Z35631	Korloy 3330	331 MPa 48 ksi	241 MPa 35 ksi	20%
Z35841	ZA27	427 MPa 58 ksi	365 MPa 53 ksi	1–6%

Extruding zinc is not common. The pressure needed is somewhat higher than aluminum. Small cross sections, rods, and solid shapes are extruded. But in general, this is rare in art and architecture. These alloys have good machining characteristics. They are extrudable in small cross sections. Available forms are:

Bar
Rod
Wire

Tubes
Forgings
Extrusion

These alloys are used in decorative applications in architectural hardware and in automotive trim. They are sometimes plated with other metals such as chromium and copper.

For the most part, they are not commonly used in architecture except possibly incorporated into hardware.

CAST ZINC ALLOYS

There are a number of methods used to cast zinc. Zinc is more versatile as a cast metal than most other architectural metals. Zinc can be cast in more ways than copper or aluminum alloys. With the exception of strength, zinc will cast as well or better than aluminum, bronze or brass. The reason for this is the low melting point and the good fluidity of the molten metal.

Temperatures necessary to melt zinc are lower, thus energy use is reduced. Zinc can be cast in steel dies repeatedly; however, the high aluminum zinc alloys will corrode steel dies. Zinc can be placed into the die using pressure, offering quick production turnaround. Zinc does not need to be degassed as other metals. With zinc, hydrogen gas in the molten metal is not a problem as it is with steel and aluminum.

In North America, one of the most dramatic uses of cast zinc in architecture was on the Legislative Assembly Building of the Northwest Territories of Canada. It was designed by the Gino Pin and the Taylor Architectural Group of Yellowknife, Northwest Territories (see Figure 2.12).

FIGURE 2.12 Cast zinc "glacier wall" on the Legislative Assembly Building of the Northwest Territories of Canada. Designed by the Gino Pin and the Taylor Architectural Group.

The zinc panels, preweathered Reinzink, were augmented with a band of cast zinc "glacial rocks" with the surface honed flat. The building was open in 1993 and in a way signaled the resurgence of zinc as an architectural metal in North America. These sand cast zinc panels are a stunning addition to the façade of the building.

Zinc sand casting is an excellent metal for creating large-scale architectural surfaces. Sand casting of zinc is simpler than casting the higher-melting-point metals such as copper alloys and steel. It is easier to arrive at a finish surface due to the softness of the metal.

Metal die casting is a common way of casting zinc. The die can be used repeatedly to cast similar parts. The molten metal enters the die by gravity, or it can be induced using pressure.

Sand casting, ceramic shell, plaster molds, centrifugal, spin casting, and vulcanized silicon mold casting can all be used with zinc.

An old method of casting zinc statues is still used to cast hollow parts, usually small parts. This process is called slush casting. The following are the two alloys used today to slush cast parts.

SLUSH CASTING

Slush casting of zinc sculpture and hollow articles of zinc came into wide use in the second half of the nineteenth century. The process of slush casting involves pouring molten metal into a mold, typically made of ceramic or plaster, and then quickly pouring the unsolidified metal back out. This in effect coats the wall of the mold with a thin layer of zinc alloy. The longer the zinc remains in the mold, the thicker the wall. The mold is then opened and the finished, hollow zinc form remains. Thousands of sculptures were made this way up until the 1950s as style and taste in art changed. In the early part of the twentieth century, aluminum was added to strengthen the zinc. The alloy was sometimes referred to as 95-5 for the 94% to 96% zinc, with the balance being aluminum. Today, smaller objects such as hollow lamp stands and statuettes are made by slush casting.

The following are alloys considered for use in slush casting today.

```
┌─────────────────┐           ┌─────────────┐
│     Z34510      │           │   ASTM B6   │
└─────────────────┘           │  ASTM B 792 │
┌─────────────────┐           └─────────────┘
│ Zinc Slush Casting │
│     Alloy A     │
└─────────────────┘
```

Alloying constituents added to Zinc

- Al: 4.50 – 5.00%
- Cu: 0.2 – 0.3%
- Cd: 0.005% max
- Pb: 0.007% max
- Fe: 0.1% max
- Sn: 0.005% max

Slush Casting 57

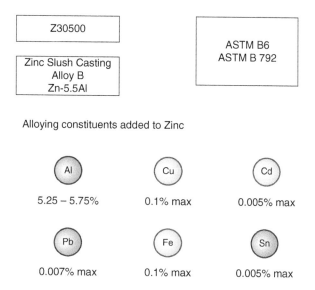

Zinc alloys Z34510 and Z30500 are ingot alloys used to slush cast. They contain aluminum, which aids in casting and provides strength. Alloy Z34510 is slightly stronger with the addition of copper. Figure 2.13 shows a slush cast replica of the Statue of Liberty. The molten zinc was poured into molds making up sections of the sculpture.

Note the welds used to join the slush cast sections together. One of the drawbacks to slush casting is you need to keep the weight manageable. Instead of attempting to slush cast the entire statue, smaller sections were cast and then joined.

FIGURE 2.13 Statuette replica of the Statue of Liberty made by slush casting zinc plates

FIGURE 2.14 Slush cast statuette. Early 1900s.

Slush casting is not used to a great degree in art or architecture today. It was widely used in the early part of twentieth century as small statuettes proliferated in gardens and parks. Many of these, such as the one shown in Figure 2.14 were commonplace. It was a decorative, repeatable process that disappeared from the vernacular of art around the beginning of World War II and never came back.

ZINC DIE CASTING

Zinc used in die casting is a significant market for the metal. The reason is the versatility, accuracy, detail, and low cost of production. These die cast alloys are designed for specific performance

TABLE 2.8 Die-cast alloys

Alloy UNS#	Common Alloy #	Aluminum	Magnesium	Copper	Iron	Lead	Cadmium	Distinguishing Attribute
Z35541	#2 Zamak 2	3.5–4.3	0.2–0.5	2.5–3.0	0.1	0.005	0.004	Highest tensile strength, good creep resistance
Z33520	#3 Zamak 3	3.5–4.3	0.2–0.5	0.25	0.1	0.005	0.004	Good strength, creep resistance and fluidity
Z35531	#5 Zamak 5	3.5–4.3	0.3–0.8	0.75–1.25	0.1	0.005	0.004	Hard alloy. Strong and good creep resistance
Z33523	#7 Zamak 7	3.5–4.3	0.03–0.05	0.25	.003	0.002	0.001	High ductility

requirements. Die casting is widespread. Zinc die castings are used in small appliances, tough housings for tools such as drill motors and saws, keys, and many other small articles that require accuracy, durability, repeatability, and corrosion resistance.

Alloy Z33520, also called alloy 3, is the most widely used die cast alloy in the United States because of the combination of attributes of fluidity, strength, and cost. Alloy Z35531, Zamak 5, is the die casting alloy favored by European die cast operations. Table 2.8 shows several of the die cast alloys in use. The common alloy number corresponds to what has been used in industry for decades. The approximate constituents of each alloy are listed.

For die cast alloys it is important to restrict the introduction of trace contaminants. Most firms that perform die casting have strict impurity limits on the metal used. Die-cast zinc parts are small, compact, and often highly detailed. They are common in many industries, just not art and architecture. Earlier in this chapter, Figure 2.4 showed a couple of examples of how small engines and housings for appliances are die cast in zinc. It is a quick and rapid process to achieve a sturdy, corrosion-resistant part with good detail.

GRAVITY CAST ALLOYS

Gravity casting of zinc has never taken hold in most foundries, limited instead to specialized firms. Gravity casting processes, sand casting, permanent mold casting, and investment casting, were concentrated in foundries that worked with copper alloys or steel. Zinc was rarely considered beyond casting Kirksite for shortrun tooling or slush casting in the early part of the twentieth century.

The International Lead Zinc Research Organization, ILZRO, in 1967, developed an alloy that contained 10.5% to 11.5% aluminum. They called this ILZRO 12, and this changed to alloy 12 for the approximate 12% aluminum it contained. It was found to have decent mechanical properties similar to some copper alloys. The benefits of the lower melting point could be exploited now with this zinc alloy. Shortly after this alloy was introduced, alloy 8 and alloy 27 were developed. Refer to Table 2.9 for the alloying constituents.

Note that the high purity alloys can be cast effectively but improved quality can be achieved with the alloys shown in Table 2.9. The approximate mechanical properties are shown in Table 2.10.

TABLE 2.9 Gravity Cast Alloys

Alloy	Common Alloy #	Aluminum (Al)	Magnesium (Mg)	Copper (Cu)	Iron (Fe)	Lead (Pb)	Cadmium (Cd)	Distinguishing Attribute
Z35636	ZA8	8.0-8.8	0.15-0.30	0.8-1.3	0.075	0.006	0.006	High strength. Excellent surface finishing.
Z35631	ZA12	10.5-11.5	0.15-0.30	0.5-1.2	0.075	0.006	0.006	Very good casting ability
Z35841	ZA27	25.0-28.0	0.01-0.02	2.0-2.5	0.075	0.006	0.006	Light weight, hard and strong. Low ductility

TABLE 2.10 Approximate Mechanical Properties of the Gravity Cast Alloys

Alloy	Common Alloy #	Tensile Strength	Yield Strength	Elongation
Z35636	ZA 8	32–40 ksi 221–276 MPa	29–30 ksi 200–207 MPa	1–2%
Z35631	ZA 12	40–50 ksi 276–345 MPa	30 ksi 207 MPa	1–3%
Z35841	ZA 27	58–62 ksi 400–427 MPa	53 ksi 365 MPa	1–6%

Alloy Z35636, also known as ZA 8 is considered for casting in permanent molds made of steel or graphite. It can be sand cast, but the result is a coarse surface. This alloy finishes well and surfacing details can be well defined.

Alloy Z35631, known commonly as ZA 12, can be cast in sand, graphite, or permanent molds. It casts very well in graphite molds. It can also be continuously cast in shapes. This alloy is used for prototyping and has good strength and wear resistance. It is a popular cast alloy due to cost. Reference Figure 2.15. Two prototype zinc "wave" panels cast in alloy Z35631.

Alloy Z35841, also known as ZA 27, is best suited for sand casting but can be permanent mold cast. It is the strongest of the cast alloys and also the lightest. It can be heat treated to improve ductility.

FIGURE 2.15   Gravity cast zinc panels. Sand cast. Alloy Z35631

KIRKSITE

Kirksite is a zinc alloy, basically similar in composition to Z35541. There are two versions, classified as Z35542 or Alloy A, and Z35543, Alloy B. In Europe it goes by the name, *Kayem*. This alloy has good hardness and toughness. This is a quick cast prototype or tooling alloy. It has good strength and is used for small part runs on press brake dies or stamping dies. Once the work is complete it can be remelted and recast. When sandcast, these alloys can achieve tensile strengths of 30–40 ksi, (210–280 MPa). An example of a kirksite die is shown in Figure 2.16. This die is tough enough to press stainless steel panels into.

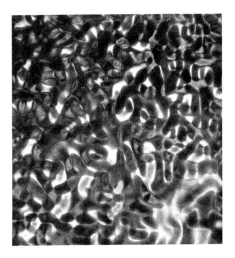

FIGURE 2.16   Kirksite "wave" die.

The die was cast, then machined to achieve a highly polished surface. The die is approximately 100 mm in thickness and has good compressive strength. The zinc acts as a lubricant allowing the stainless steel parts to slip over as they are formed.

Short run press dies are often made from this alloy. Cast then machined, these dies enable prototype runs and unique parts without extensive cost of production tooling.

CHAPTER 3

Finishes

During the course of my research, I had had occasion to examine not only simple compounds, salts and oxides, but also a great number of minerals.

Source: Marie Curie, Pierre Curie with Autobiographical Notes. Translated by Charlotte and Vernon Kellogg (New York: Macmillan, 1923)

INTRODUCTION

When we think of zinc, the dull, low reflective gray metal that wraps much of the Paris skyline comes to mind. Or perhaps the spangled, white gray metal surface of galvanizing on steel is what we associate with the metal zinc. Zinc is actually more common than we first realize. Today it is used more frequently as an architectural cladding material comparable to copper and painted aluminum.

Figure 3.1 is a thin natural zinc taking on the appearance of silver clay.

Zinc usage in North America does not have the deep history it has enjoyed in Europe. In Europe resides some of the largest manufacturers of zinc products and zinc is a common architectural metal considered for roofing, cladding, guttering, and ornamentation. In North America, the designer is less familiar with zinc, as well as the sheet metal industry tasked with working with the metal. The lack of knowledge can lead to misuse and frustration.

Much of the information on working with zinc has come from German and French producers and has been adopted by North American companies. The knowledge of the metal and how to best work with it needs further understanding to those firms involved with designing and building with the metal. There are differences between zinc, aluminum, copper, and stainless steel that need to be understood and planned for.

64 Chapter 3 Finishes

FIGURE 3.1 Zahner Engineering space. Natural zinc, clay-like appearance. Designed by Crawford Architects.

FIGURE 3.2 Zahner Engineering space. Dark and stormy day.

The beauty and behavior of the zinc surface is unlike any of the other metals used in art and architecture. It is a natural material, an element, not unlike aluminum and copper in the sense that it has been rendered from minerals to a level of considerable purity. Because of the energy that went into refining zinc, there is the constant drive to join with the other elements in our environment. Zinc combines rapidly with oxygen and carbon dioxide, trapping this greenhouse gas permanently in a protective compound on the surface of the metal.

Zinc changes in appearance as it interacts with the light of the day and the conditions of the environment.

New patination techniques and enhancements to the surface of zinc are only recently being explored and introduced to the art and architectural marketplace. This "old world" metal is experiencing a rebirth as an architectural metal in North America. Designers are just now beginning to embrace the beauty and nature of zinc. The working knowledge needed to construct with the various forms of the metal, are advancing. Casting and texturing of the zinc surface are finding new vigor. The future of the metal as an architectural and artistic surfacing material beyond roofing and guttering is coming to form.

This chapter discusses the common finishes produced and available for zinc and a few of the not so common finishes as well as several experimental zinc surfaces. Table 3.1 lists the various zinc surfaces discussed.

APPEARANCE AMONG METALS

From an architectural context zinc falls in the middle between the metals that are not expected to change, such as stainless steel and aluminum, and the metals we do expect to change, copper alloys and weathering steel. The zinc surface changes but very slowly as it combines and absorbs substances in our atmosphere.

Zinc is rarely coated with organic films to protect it from the environment. There are some manufacturers of the metal applying thin tinted acrylic coatings to adjust the color. Zinc will receive most organic paints well with the exceptions of oil based paints.

Typically, zinc is expected to weather, only very slowly and predictably. Like the roofs in Paris, ranging from 300 year old surfaces to those newly installed surfaces, on general examination you can discern the new from the old, but it all blends together quite well. The oxide that forms over time is not thick and the preweathered new zinc surface is similar in appearance. The oxide that forms over years of exposure is zinc carbonate. Zinc pulls carbon dioxide out of the surrounding air and captures it as an inert, stable layer on the surface.

Zinc is one of the silver tone metals. Most metals are silver or gray. Copper and gold are the exceptions. Zinc is difficult to distinguish from the other metals when in the natural, unfinished state, see Figure 3.3. Each of the samples shown were given a light satin finish.

Figure 3.4 shows the relative reflectivity of several metals. The graph of zinc represents, natural, unoxidized zinc without a pre-weathering. The graph indicates zinc reflects the wavelength of visible light similar to silver on the red wavelengths but more like stainless steel on the blue wavelengths. This gives zinc a slightly bluish cast. Most zinc used in art and architecture is provided in a preweathered condition. That is, there is a chemical compound of zinc and carbonate or zinc and

TABLE 3.1 Sources of Various Zinc Finishes

Finish	Comment
Natural finish – as rolled sheet	Produced by several mills
Natural finish – angel hair	Secondary fabricator
Natural finish – glass bead	Secondary fabricator
Natural finish – plate zinc	Produced by several mills
Natural finish – as cast – sand	Secondary fabricator
Natural finish – as cast – smooth	Secondary fabricator
Mechanical finish – embossed	Secondary fabricator
Preweathered – zinc carbonate	Produced by one major mill
Preweathered – zinc phosphate	Produced by several major mills
Preweathered – pigmented	Produced by several mills
Blackened	Produced by several mills
Patinated – Roano™	Secondary fabricator
Patinated – Baroque™	Secondary fabricator
Patinated – Hunter™	Secondary fabricator
Patinated – Custom/Experimental	Secondary fabricator
Copper plate	Secondary fabricator
Custom – melted	Secondary fabricator
Patina – iridescent	Secondary fabricator
Galvanized – spangle	Secondary fabricator on steel
Galvanized – darkened	Secondary fabricator
Galvanized – zinc phosphate	Secondary fabricator
Galvanized – zinc fabric	Secondary fabricator
Other zinc coating processes on steel	Secondary fabricator
Zinc anodizing	No known processor for architectural use

phosphate that has formed on the surface and is an integral part of the metal. These are thin coatings that stabilize the initial surface and provide a consistent finish appearance.

If natural zinc was allowed to age, it would lose the reflectivity and darken as it forms the zinc carbonate by absorbing and interacting with moisture and carbon dioxide. The graph would be similar but the reflectiveness as a whole would decrease. The blue-gray color tone would be more pronounced because the surface is diffuse and light scattering rather than point to point reflectivity.

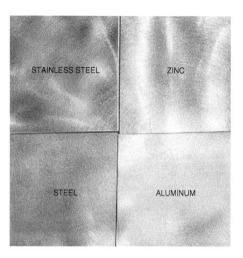

FIGURE 3.3   Stainless steel, zinc, aluminum, and steel.

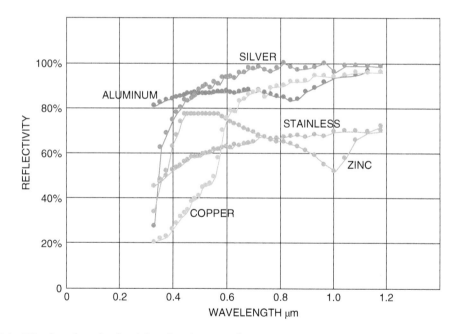

FIGURE 3.4   Wavelength and reflectivity of various metals.

Figure 3.5 depicts what happens when light scatters as it reflects off a surface as compared to a highly reflective surface. Therefore, preweathered zinc is a nonglare surface.

From an art and architecture prospective, zinc is available from the mill source in sheet, plate, and ingot form. Zinc is available from secondary fabricators and manufacturers in other forms.

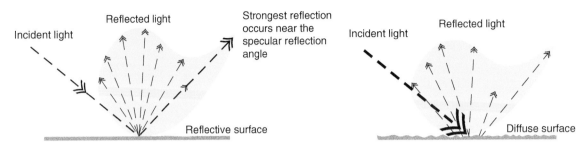

FIGURE 3.5 Reflective surface as opposed to a diffuse surface.

Galvanizing of steel, for example, is performed by a secondary facility that acquires the zinc ingots and melts them to create a molten bath of metal for steel.

Similarly, the numerous companies that die-cast zinc obtain zinc alloys with specific content and melt the metal down for processing.

Patinated, textured, and expanded zinc are available from firms that take the rolled zinc mill forms and convert the surface further to enhance the metal for a particular application or order. Zinc all starts at the mill source.

MILL FINISHES

The mill source creates natural, as cast or as rolled zinc alloy. Other than ingots, which are simply blocks of solid metal to be reprocessed, the mills will provide continuously cast plate or sheet in a natural, as fabricated form. The mill producers will also take the sheet and coil and preweather the surface to enhance the appearance and improve the corrosion resistance.

From a business viewpoint, most zinc mill producers are more vertically integrated with their value chain, they cast, they preweather, even manufacture product. This is somewhat different than the producers of the other metals used in art and architecture. It harkens back to the early days of aluminum where a major mill producer may provide a brand of fabricated roof panels or wall panels.

NATURAL ZINC COLOR

The natural zinc color when first cast or continuously cast and rolled into sheets is a reflective, silver color. Kept dry, it retains this color, but wet from rain or condensation, humidity or handled excessively, the surface darkens and loses its reflective sheen after a short period of time. Zinc tarnishes, and the surface roughens as it develops zinc oxide from exposure to the air, particularly humid air.

The surface of natural zinc when cast in sand, refer to Figure 3.6, takes on the sand texture, however when cast in smooth walled steel or ceramic dies, the surface of the cast zinc takes on this smoothness as shown in the right image of Figure 3.6. Molten zinc flows well and the early use by

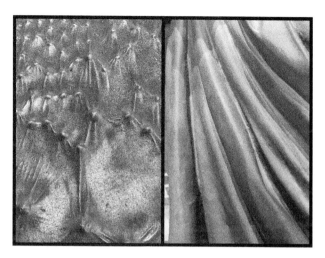

FIGURE 3.6 Sand cast zinc surface and ceramic cast zinc surface.

artists took advantage of this by casting zinc in smooth wall molds to achieve a good surface finish with little work.

The die-cast alloys produce finer grains. The die-cast alloys contain aluminum and the finish surface is very bright and smooth. These surfaces can be plated with copper or chromium to create household fixtures or lamp features. The surface can also be left to age. The aluminum zinc alloys have excellent corrosion resistance and the cast surface remains free of whitish compounds of zinc oxide. The size and scale of the die-cast product is a limiting factor for use in art and architecture.

For architectural surfaces where the rolled sheet is used as cladding for walls or roofing, the natural, nonweathered surface is rarely used. Those first sheets of zinc rolled for the roofs of Paris in the early part of the nineteenth century must have shimmered in the sunlight. These were not preweathered but were most likely rolled hot and had a layer of oxide that gave them a darker tone. Still the distinctive color of these roofs continues to present time as the mansard roofs blend nicely with the blue-gray sky.

Thin, natural, as-rolled zinc sheet when installed on a surface will appear rippled and distorted as the thermal expansion and contraction move the metal. Internal differential stresses inherent in thin sheet cause some local instability as the surface heats up. Commonly known as *oil-canning*, the higher reflective surface of new rolled sheets, emphasize the most minor differences of the out of plane diaphragm. As the surface weathers, this reflectivity diminishes, and the distortions are less pronounced. They are still there, just not enhancing any longer by a reflective surface. It takes time for the oxide to develop and tarnish to dull the reflective surface.

Plate thicknesses of zinc are available in only natural color tones from most zinc mill producers. Secondary facilities that specialize in patination of zinc can preweather the surfaces to darken and even the appearance out. The difference with the thicker plates is in the surface texture. Zinc rolled at thicknesses of 4.5 mm and heavier are considered plate and their surfaces are coarse and less refined compared to the thinner cold rolled sheet material.

FIGURE 3.7 Zinc planters.

Figure 3.7 shows the surface of a zinc planter made from 6 mm thick zinc plate. The surface texture is a hot-rolled, coarse surface. The coarseness adds to a rugged natural appearance. The white streaks on the surface are zinc hydroxide. Zinc hydroxide is very adherent and stable, making it difficult to remove. It develops on the zinc surface from moisture collecting on the surface. The mottled color tone adds to the natural appearance. These zinc plates should perform and last a very long time.

The surface has a different character than hot-dipped galvanized steel plates. It has the color but lacks the spangle of the crystal. Continuous wetting of the surface will not damage the inner zinc particularly at this thickness. These will continue to weather. Scratches will not harm the plate because there the oxide will just return, and the metal is consistent throughout.

Corrosion products from zinc are clear and do not stain adjacent surfaces. This is one of the attributes that make natural zinc particularly useful as a cladding material. Similar to aluminum, its oxide does not pose a staining issue. Other substances corroding near or above natural zinc surfaces will stain the zinc, however, and these compounds often will combine with the zinc oxide and hydroxide on the surface. It is wise to not have copper runoff falling on zinc surfaces nor allow steel rust to run onto the zinc surface.

MECHANICAL FINISHES

The mechanical finishes are imparted to a metal surface by means of physical interaction. With architectural metal this usually involves sanding belts or abrasive disks, glass beads and energy. The metal on the surface is altered as higher energy combined with abrasive substances impart scratches and tiny craters on the surface. In the case of a highly reflect zinc surfaces, this could entail very minute scratches and sometimes electrochemical dissolution to provide a texture.

Unlike the other metals used in architecture, stainless steel, aluminum and copper alloys, mechanical finishes applied to zinc surfaces are not common. Zinc can receive mechanical finishes. These finishes expose the natural, unoxidized zinc. Surface refinishing of table or bar tops are cleaned frequently with mild abrasives. But as a general offering, zinc sheet does not typically receive a linear satin finish or nondirectional finish. Part of the reason lies in the rapid tarnish that will development on the natural zinc and the affinity for fingerprinting of the natural surface. See Figure 3.8 for examples of several mechanical finishes and rolled textures applied to zinc sheet. These textures were induced onto the preweathered surfaces, which gain an interesting contrast as the darker preweathered layer has been scratched through to expose the underlying, more reflective natural zinc.

Even a mirror polish can be achieved, but the finish is short-lived unless it is protected with a clear coating. When polishing zinc, the surface tends to smear on a micro level. You achieve high and low points that induce a mottled appearance; some regions reflect light at slightly different levels and appear slightly darker than other areas on the surface. The natural surface will fingerprint and tarnish rapidly. Fingerprints in zinc will etch the surface and require a tarnish remover to restore the surface. This is one reason why most zinc sheet material is provided with a preweathered zinc carbonate or zinc phosphate coating. These resist tarnish and fingerprinting and protect the surface from minor contaminants.

Cast zinc can take impact from abrasive blasting. Sand cast zinc often needs to be lightly blasted to remove mold residue. This induces a bright, diffuse texture initially, but as exposure to air and humidity this dulls down to a soft pewter-like appearance.

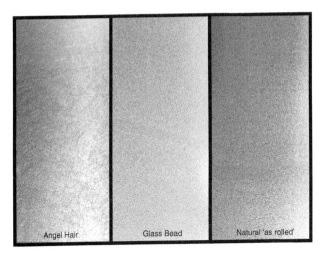

FIGURE 3.8   Mechanical finish examples on zinc.

MECHANICALLY ROLLED TEXTURES

Rolled textures are available on zinc. The sheets or coils of zinc can be processed through a set of rolls that impart a light texture to the surface. Zinc accepts the rolled finishes well as pressure rolls induce a negative texture into the sheet zinc material. The metal yields and conforms to the imprinted roll surface. Figure 3.9 shows several examples of textures induced into preweathered zinc sheet.

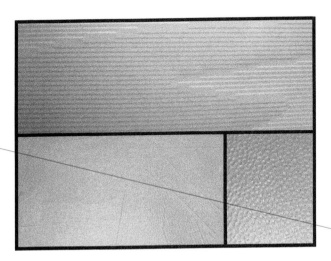

FIGURE 3.9   Textured zinc.

Texturing the surface does not harm the preweathered surface coating. The surface is flexible and integral to the zinc. It will not flake or crack from adding texture to the surface.

Textures applied this way stiffen the zinc surface allowing thinner zinc sheet to be used or adding a level of durability to resist denting.

PREWEATHERED ZINC SURFACE

Zinc used in art and architecture is often preweathered or preoxidized at the mill source. The preweathering process is a surface treatment that creates a stable, corrosion-resistant transition film. The treatment involves either a phosphoric acid surface treatment or a sulfuric acid treatment.

In the phosphoric acid treatment, the natural rolled zinc coil is placed in a phosphoric acid bath to form zinc phosphate and hydrogen. The result is an insoluble, tertiary zinc phosphate compound, $Zn_3(PO_4)_2$.

In the sulfuric acid treatment, the commercial pure zinc coil is treated with sulfuric acid to form zinc sulphate and hydrogen. This is further treated to a bath of sodium carbonate to form zinc carbonate over the zinc surface and sodium sulfate in solution. This is performed at the mill source of rolled sheet zinc. The zinc carbonate is thin but very similar to what forms over zinc when exposed to the atmosphere. The formula is $ZnCO_3$.

Preweathering allows the mill to store the zinc in coil for further processing or inventory, reducing the concern for damage from moisture or tarnish.

Depending on the mill, the preweathered surface is provided in one of two forms, zinc carbonate or zinc phosphate. Rolled zinc sheet supplied from one mill will not match rolled zinc supplied from a different mill. There are subtle surface variations from mill to mill as the equipment, processing and preweathering operations are unique to the specific company.

The zinc carbonate preweathering surface has a different color tone and reflectivity than the zinc phosphate. Figure 3.10 a preweathered zinc surface with the zinc carbonate finish. Not all producers of zinc sheet provide this surface finish. Most provide the zinc phosphate preweathering surface.

Preweathering offers several benefits to the thin-rolled zinc strip and sheet. First of which is protection during handling, forming, and shipping. The preweathered surface offers excellent protection to the zinc surface as the metal is processed. Fingerprints, moisture, and most oils wipe off the surface without interacting with the preweathered layer.

Preweathering leaves the surface with an even, low-matte gray tone. The gray tone may have a hint of green or blue, depending on the preweathering process used. The zinc carbonate combined with rolling on polished rolls gives a slight greenish hue. Figure 3.11 shows the initial difference between the zinc carbonate and the phosphate preweathered surface.

The color and the surface preweathered layer are a natural extension of the zinc alloy. It is not applied as a separate coating like a paint, it is a chemical reaction that develops a very thin layer over the surface. The surface is very adherent and will not flake or crack when thin zinc sheet is folded.

FIGURE 3.10 Preweathered zinc. Zinc carbonate preweathering on 2 mm thick zinc plate. Institute for Contemporary Art at the Virginia Commonwealth University. Steven Holl Architects.

FIGURE 3.11 Comparison of the zinc carbonate (left) and zinc phosphate (right) preweathered surfaces.

Unlike other metals, zinc, as it weathers and the surface grows, does not generate runoff capable of staining adjacent materials. Any runoff that is generated as the oxide grows is usually light in color. In urban environments zinc sulphate or zinc sulfite can develop and these are water soluble. These compounds are light in color. These can develop when the preweathered zinc carbonate or the zinc oxide is attacked by sulfur compounds found in industrial environments.

The most important aspect of the preweathering, is the corrosion resistance this surface compound offers as the zinc is exposed to the environment. If properly applied it greatly extends the service life of the metal and for most environments where zinc is used as a cladding, this results in a very slow change in surface appearance. The main exception is when the zinc is used in a polluted environment and the water soluble sulfite or sulphate forms.

In coastal environments the surface lightens from the development of zinc chloride on the surface. Zinc chloride is partially soluble in water and will collect in drip areas and slow draining zones.

Rolled, preweathered zinc sheet, has a directional grain that is more pronounced than one might expect. It is actually very similar to rolled aluminum alloy sheet. The grains are apparent at close inspection and can have an initial influence on appearance if elements are constructed and set at angles to the grain direction. The grain is more apparent in new sheet material versus material that has been exposed for a length of time. Figure 3.12 shows the grain of a preweathered sheet. The grain direction is pronounced in rolled zinc sheet and plate, specifically the copper–titanium alloys.

It is also important to understand that different batches from the same manufacturer processed at different times can lead to differences in the preweathering tone. The process may be the same, but the degree of surface development can be different, leading to a patchwork of tonal differences. This can also occur if the grain direction is not followed or if the sheets are flipped in the process of producing the panel. Figure 3.14 shows this affect. The effect is minor on this stunning example of architecture. The variations adds to the feeling of looking at a natural surface rather than an artificial surface.

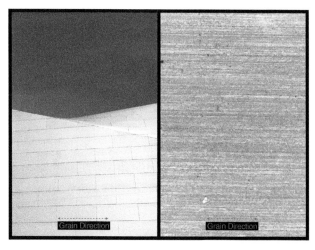

FIGURE 3.12   Grain direction of rolled zinc.

Most, if not all, of the major producers of rolled zinc sheet provide a preweathered surface, often with enhanced choices of color and tone. There are light gray, blue gray, black, and tinted tones. The preweathering is performed on all exposed surfaces of the sheet or plate. You should be aware that not all preweathered surfaces are identical. One manufacturer will process the metal coil or sheets differently than the next. Similar to anodized aluminum, you can expect the color tones created from the preweathering process used by one firm to be different in appearance from that of another. Figure 3.13 shows a few examples of this.

Chapter 3 Finishes

FIGURE 3.13 Preweathering of similar thicknesses of zinc performed by different firms.

FIGURE 3.14 Thick zinc sheets showing a slight color variation on initial installation. Institute for Contemporary Art at Virginia Commonwealth. Designed by Steven Holl Architects.

> When working with preweathered zinc, it is important to recognize:
>
> - Preweathering is a chemical transition of the outer zinc surface.
> - Preweathering develops a surface similar to a naturally developing surface.
> - Preweathering produces a stable, passive surface.

- Rolled, preweathered sheets from different manufacturers will appear different.
- Rolled, preweathered sheets processed at different times will appear different.
- In most normal environments, preweathering will slow reactions of the surface.
- The oxide on zinc is light in color and will not stain adjacent materials.

Unlike the prepatination of copper or the preweathering of weathering steel, the oxide created on zinc is very adherent and will not stain adjacent surfaces. Further weathering is very slow to occur. There may be areas where a white, friable substance appears where water collects and stands. This is zinc hydroxide that develops as zinc ions migrate through to the moisture. On well-drained surfaces, this does not occur as long as it is allowed to dry completely. In marine environments, zinc chloride will develop, and this also has a white appearance.

Many roof surfaces clad in zinc accommodate a space just below the underside of the thin sheet to allow for air to flow and facilitate drying. A period of prolonged exposure to moisture on the zinc can develop into corrosion as other substances combine with the moisture to create localized corrosion cells. If sulfur is present, the corrosion product of zinc sulphate or zinc sulfite will sluff off the surface, exposing more zinc to repeat the process.

CLEAR COATING WITH PIGMENTATION

There are several companies that produce zinc sheet with a slight color tone induced. Red, blue, green or olive color and bronze colors are induced by adding a pigment to a clear lacquer. The lacquer coating is only 10–12 μm in thickness. This imparts a color to the surface as light passes through the clear coating and reflects off of the surface, the pigment gives a slight color tone to the metal.

The coatings add to the corrosion resistance of the zinc and appear to perform well in ultraviolet. They take away the character of the zinc and even out the appearance to the point they could be any metal with a paint coating. Figure 3.15 shows a project with a blue pigment introduced to the clear coating on the zinc.

BLACKENED ZINC

Blackened zinc is available from several manufacturers. This is an extension of the preweathering process. Blackening of the zinc surface is a darkened conversion coating. It can be induced on the surface in several ways. A zinc phosphate darkening treatment, a chlorate darkening treatment, even a copper selenide treatment can create this conversion coating with a deep matte color. There are several commercial production processes performed by the suppliers of the metal that make the black zinc available.

78 Chapter 3 Finishes

FIGURE 3.15 Tinted zinc coating.

FIGURE 3.16 Blackened perforated and bumped zinc used as a bridge cladding. Designed by Helix Architects.

Figure 3.16 shows a blackened zinc wall cladding for a utility bridge. The surface was custom perforated and embossed with the "+" sign or the "−" to depict the electrical utility being housed in the enclosure. The black color is very stable and adherent. The surface is matte black. Over time, it will age to a dark charcoal color as regions of zinc hydroxide form.

Similar to the preweathered surface, different batches of blackened zinc will have different shades. Like the preweathered surface, this black color is created by a chemical reaction with the zinc. It is not paint or a coating that can be easily delineated from the base zinc material. The zinc color is a matte black color tone.

CUSTOM PATINA FINISH

Zinc is never found in its native (pure), state in nature. It is always rooted in combination with other elements usually in complex mixtures of other metals, in particular lead, iron, and cadmium, along with oxides, sulfides, and chlorides depending on the exposures.

As shown by the atomic structure of zinc in Chapter 1 and the thermodynamic nature of the zinc ion described in the formula:

$$Zn \leftrightarrow Zn^{2+} + 2e^- \quad E_0 = -0.763v$$

Zinc will quickly join with oxygen to form the oxide and hydroxide when moisture is present. Zinc will also rapidly join with other elements and compounds to form other insoluble compounds on the surface. This explains why the preweathering compounds of zinc phosphate and zinc carbonate are two very adherent, corrosion resistant compounds that resist change. When produced they act as thin mineral-like barriers. The carbonates were also used to develop the tone and color on many of the early slush cast sculptures to arrive at the beautiful, pewter-like appearance of zinc carbonate. Figure 3.17 shows an example of such a sculpture. Zinc carbonate will form naturally on the outside surface of zinc exposed to the atmosphere as carbon dioxide is absorbed and exchanges with the zinc hydroxide on the surface.

FIGURE 3.17   Zinc sculpture. Kansas City Plaza.
Source: Courtesy of Zahner Metal Conservation

As an oxide forms on the surface of zinc, a barrier is created. Ionized atoms of zinc must pass through this barrier for the oxide to continue to develop and grow. At first, the oxide grows quickly until a thickness is achieved that slows the diffusion of ions way down.

DARK VARIEGATED PATINAS ON ZINC

This ability to quickly and effectively form insoluble compounds is the basis of the patinas that can be developed on zinc artificially by introducing compounds that bond with zinc. The process involves activating a flow of Zn^{2+} ions into a reactive solution. The solution is an acidic electrolyte containing negatively charged anions that combine with the zinc at the surface to form mineral-like natural compounds. The compounds must join with the surface slowly to arrive at a conversion layer of insoluble, yet adherent compounds. Figure 3.18 shows a finish known as *Roano*™.

This patina is a stable, highly variegated surface. As exposure increases the frequency of moisture within mild light industrial conditions, the patina will gradually change. Usually a light white spotting of zinc hydroxide appears as the moisture is held by the semiporous surface or as moisture collects on drip edges or flat regions. Change in base color is minor. Figure 3.19 is the Taubman Museum of Art. The Roano™ finish was applied to this structure over 10 years ago. The surface, particularly around the edges has some minor zinc hydroxide whitening, otherwise the surface looks very consistent with the day it was installed.

The finish has an interesting directionality due to the depth of the crystalline surface that forms. The crystalline surface as shown in Figure 3.20 is rough, but continuous with zones of heavy crystalline development and areas of openness. The green hue is generated from the optical profile scan

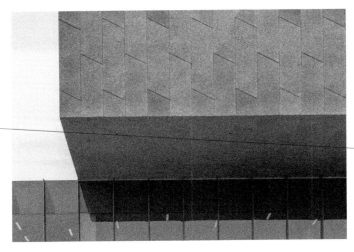

FIGURE 3.18 Roano™ patina on rolled zinc alloy Z41121

FIGURE 3.19 Taubman Museum after 10 years of exposure.

used to capture the surface image. The surface is similar in nature to the mineral franklinite, a zinc iron oxide, and the rare mineral known as *stelleite*.

This amazing micro-topology traps and scatters light similar to the way a natural mineral surface does. The appearance, color, and tone of the reflected light will change slightly as the angle of view changes. This is also why there is an appearance difference from panel to panel. The surface "grows" as the zinc reacts and thickened corrosion products form. Figure 3.21 is the Roano™ patina created for a lobby wall.

The process is similar to prepatinated copper or preweathering steel. The surface goes through a chemical reaction as some of the zinc on the surface goes into solution and combines with other elements in the solution. These reactions must occur at the surface and time is needed for zinc ions to diffuse and create bonds with the elements in the patina solution. Like all chemical reactions their

FIGURE 3.20 Microscopic view of the Roano™ patina on rolled zinc.

FIGURE 3.21 Interior surface using Roana Zinc patina.

efficiency is dependent on time of exposure and temperature of the metal. If the process is rushed, you can get separation of the finish. If the temperature is below 20°C (68°F) the chemical reaction is slow to occur as is the case with other patination processes on copper or steel.

For this particular finish, called Roano™ zinc, there are a series of reactions that occur on surface in a layering effect. As the finish develops an appearance not unlike the minerals that give some paving stones color, forms a tight bond with the zinc. Figure 3.22 shows the Taubman Museum compared to a stone flooring. The lower righthand corner is the stone flooring material.

FIGURE 3.22 Taubman Museum surface compared to stone flooring with similar mineral color characteristics.

A variation of the Roano™ finish is called Baroque™. With this patina the variegated finish is darkened, still very stable but with a darkened zinc background. The darker tone adds richness to the design. The chocolatier company, Max Brenner™, used this finish to emulate the deep color of the product it produces. See Figure 3.23.

There are other variations on this finish produced by layering of the chemistry not unlike the layering of patinas on copper alloys. In a similar fashion, each layer must adhere, and this is accomplished when the chemical reaction occurs near or at the surface and has sufficient time for diffusion of zinc out from the base metal into the patina layer.

FIGURE 3.23 Chocolate company Max Brenner™. Clad with shingles of Baroque patina zinc.

Developing patinas is one of the beauties of zinc that have never been fully exploited commercially as they have with copper alloys. Copper and its alloys would develop beautiful colors by combination of the various copper salts of sulfides and chlorides. It was an artform that has roots over centuries of patination on bronze statues.

Zinc statues and statuettes were popular soon after the production of zinc in quantities occurred in the early 1800s and the French and Germans advanced this artform as they decorated their cities. Zinc was known as "white bronze" during this period and on into the early 1900s. Applying a patina was not attempted to any great extent beyond blackening the surface.

Plating these sculptures with copper became a common method of adding color and, perhaps, increasing the marketability. In 1869 the process of plating zinc with copper became common for adding panache to the zinc sculpture. This carried into the United States into the 1930s as many sculptures made of zinc were plated with copper. Figure 3.24 shows two examples of zinc statues that are plated in copper. These post–World War I sculptures are hollow cast zinc and plated with copper to give them a bronze tone. After nearly 100 years they still are in good condition. There are some cracks due to creep as the zinc elongated under the constant load of gravity and the environment.

Casting with zinc to produce large art forms is not as common a practice as it once was. Plating of large sections of zinc is not common either. The art is not lost, however, and plating with copper is performed several million times a week to create the penny. The penny after 1984 is stamped out of zinc blanks and plated with copper. Millions of blanks are produced, and mass plated with copper each week in an amazing production process.

Zinc and its reactivity afford intriguing color tones that are an extension of the stable mineral surfaces that are found in nature. The key is to arrive at insoluble and tightly bound patina.

FIGURE 3.24 Cast zinc statues of WWI solders copper plated to mimic bronze.
Source: Courtesy of Zahner Metal Conservation

FIGURE 3.25 Children's Hospital of Richmond Pavilion. VCU Campus. Designed by HKS Architects.

Figure 3.25 shows the parking garage at the Children's Hospital of Richmond Pavilion, on the VCU Campus. The designer, HKS, wanted to create earth tones using the Roano™ finish patina process as a base. Several variations of the finish were designed into the parking garage façade to add color to the texture by layering patinas on some of the sheets. Others were blackened. Each of these finishes is a patina induced onto the zinc by creating a slightly different compound on the surface. These were not sealed or coated. The patina will age gracefully over time and exposure. The differences in color are mineral in form and will not wear away or fade.

Not unlike the patina on copper or weathering steel, these zinc patinas take time to develop. Each is produced in a controlled environment that enables the zinc ions to diffuse into a solution and react with other ions and form a thin adherent patina on the surface. The patina that develops is insoluble and bonds to the base metal. Figure 3.26 and Figure 3.27 are the Taubman Museum of Art in Roanoke, Virginia. The Roano™ patina was originally developed for this project, designed by the late Randall Stout. He wanted a surface that related to the natural fall colors of the Blue Ridge mountains that surround the city.

It is critical to understand the patina that grows on zinc like that on copper or weathering steel is a chemical reaction. The ability to precisely match one sheet to another is equivalent to searching for pebbles in a stream that look exactly alike. You will get some close but never an exact match. However, the end result should be within a set family of color tones.

There have been instances when a range is required to be established. Producing range samples are an attempt to objectify whether something is acceptable or not. Such attempts to express a condition that is more abstract into concrete terms can lead to frustration. It is best to arrive at agreeable and controllable parameters that can actually be achieved.

86 Chapter 3 Finishes

FIGURE 3.26 Taubman Museum of Art. Roanoke, Virginia. Designed by Randall Stout.

FIGURE 3.27 Roano™ Zinc on the Taubman Museum of Art.

To Manage Expectations:

- Produce control samples from the actual metal to be used – establish a direction arrow.
- Establish a density of layering – have control samples to compare to.
- Apply the patina in a controlled environment – the process should not be random.
- Control temperature, humidity, pH – establish measures.
- Allow ample time for the patina to bond – do not force things to occur.
- Avoid groupings of light tones and dark tones together – exercise installation controls.

FIGURE 3.28 Polk County Criminal Court Building.

Since the process requires the patina to develop over time while stored in a controlled environment the stage of verifying whether a surface falls into an acceptable color or tone is critical. The finish on day one is going to be a little different than the finish on day five.

Figure 3.28 is an example of the expected variation in the Roano™ patina. Numerous panels were made from sheets of Roano™ finish zinc. Each panel is unique, but the patina is very natural in appearance and the surface mosaic created by slight differences adds to the appearance.

The variations are part of the beauty with this finish. It has a natural look and is pleasing to the eye. It is a mineral surface and the variations are appealing.

ZINC OXIDE PATINAS

Another patina, developed in a similar fashion, but somewhat simpler in make-up, uses the oxides of zinc to create an adherent marbled surface. This finish goes by the trade name Hunter™. Figure 3.29 shows the Hunter Museum of Art in Chattanooga, the first project to use the finish. The Hunter Museum was designed by Randall Stout. He was struggling with an economical surface material that would fit well with the dolomite cliffs of the Tennessee River. The museum was to be constructed on top of the cliffs overlooking the river. Randall Stout wanted a surface that looked as if it was formed out of the rock cliffs. The museum is perched directly on top of the cliffs.

The finish had been under development for several years. It all began with a restoration project. The author was engaged in the restoration of a Donald Judd art piece that had been damaged by exposure to moisture while in storage in London. The art piece was a series of large spatial forms in the shape of a cube. They were made from thin hot-dipped galvanized steel.

The galvanized surface had developed a thick, zinc hydroxide and zinc oxide crust over sections of the surface. Those of us that work with sheet metal inevitably experience the effects of moisture

FIGURE 3.29   Hunter Museum of Art. Chattanooga, Tennessee. Designed by Randall Stout.

FIGURE 3.30   Thick, white zinc hydroxide formation on a Donald Judd sculpture.

entering between stacked or coils or sheets of galvanized steel. The resulting stain, a whitish crust, is tenacious and near impossible to remove without damaging the underlying zinc. See Figure 3.30.

Several months were spent working with methods to remove the stain from the artwork without harming what remained of the galvanized surface. The stain is superficial and does no harm other than an appearance challenge. It actually adds a level of corrosion resistance to the surface. From the work on the sculpture, an understanding of the tough, inert nature of zinc compounds and how, once they are present, removal by any normal means is futile. The zinc hydroxide was resistant to generally all mild solvents, acids, and bases. One could conjecture if this could be reproduced, it would resist the milder effects presented by the natural environment and last indefinitely.

FIGURE 3.31 Hunter™ zinc on the Hunter Museum of Art.

The base surface of the Hunter™ zinc is a preweathered surface. See Figure 3.31. The zinc hydroxide is "grown" onto and out of this base surface. The effect is a natural-looking, mottled color not unlike the dolomite cliffs the museum is built over. Dolomite, the mineral, has no zinc in its makeup; however, the mineral has a similar, hexagonal crystal structure like zinc. Dolomite is made of calcium magnesium carbonate and has a white marbling over a gray base, much like the marbling created on the preweathered zinc to produce Hunter™.

The Hunter™ oxide surface has variations induced onto the surface. The marbling is not unlike the natural development of streaks in minerals, as concentrations over time cause different compounds to develop. The zinc hydroxide must have time to develop on the surface as zinc ions diffuse from the surface and react with oxygen in the electrolyte to form into the oxide and hydroxide. The result is the white, thickened outgrowth from the zinc surface. This roughened white substance adheres to the underlying zinc. As it develops, differing thicknesses and intensities create variations on the surface of the darker zinc. In some cases, other compounds in the electrolyte form and the streaks can go dark in contrast to the lighter oxide. Figure 3.32 shows where dark and light tones are imparted to the same zinc sheet surface.

Processes of application can alter the exposure of the zinc surface to create the mottling in a selective manner. Once the oxide forms, it slows the further development of the patina down. This selective corrosion of the surface is neutralized to stabilize and limit further oxidation. The right image on Figure 3.32 is an interior surface of the Hunter™ zinc patina. On interior applications, the surface is very stable and changes in appearance over time will be negligible.

This finish can develop on formed panels or on sheets. The finish is not flexible but it is thin and very adherent. Postbending, the surface may create microfractures, but these are hardly visible. The zinc hydroxide will not flake or rub off the surface.

Figure 3.33 is the Art Gallery of Alberta, also designed by Randall Stout. The exterior surfaces of the gallery spaces were clad in Hunter™ zinc. The panels in this case were first formed and then finished.

FIGURE 3.32 Hunter™ zinc with different tones.

FIGURE 3.33 Hunter™ zinc on the Art Gallery of Alberta. Designed by Randall Stout.

On the Art Gallery of Alberta, the panels were made into trapezoids, keeping a "water-line" horizontal, while breaking the vertical line. Each panel was patinated separately to further enhance the texture of the overall zinc surface. Figure 3.34 shows the effect of this.

Note, when patinating formed panels you will get a halo effect at the folds due to surface tension and drying patterns. This tends to enhance the vertical edge. This difference remains as the metal continues to age. The darker background lightens only slightly as the zinc hydroxide forms, then darkens as carbon dioxide is absorbed to create the ultimate zinc carbonate surface.

Zinc Oxide Patinas 91

FIGURE 3.34 Art Gallery of Alberta. Custom Trapezoid panels. Designed by Randall Stout Architects

There are other patinas that can develop on zinc, each with various levels of durability. As discussed previously, zinc is never found in nature in the native, pure state. Zinc is so intertwined into various minerals because of its drive to combine and form compounds. Patination can take advantage of this by introducing to the surface compounds that once form and stabilize into a barrier of enhanced color.

Figure 3.35 shows a few patinas that are similar to the Roano™ and Hunter™ patinas and some significantly different. Each of these is produced by chemical reaction with the zinc and zinc oxide on the surface.

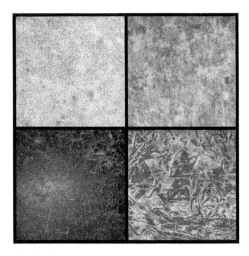

FIGURE 3.35 Other zinc patinas in development.

Producing these patinas require a clean, oil-free surface. Some of the zinc sheet manufacturers treat the product with a light protective oil or thin organic film. This must be removed in its entirety for patination to work.

As with copper, developing the patina requires time for the chemical reaction to occur. The zinc surface is homogeneous, and once degreased and clean, the reaction forms over all the exposed surface of the sheet. It should be understood when developing patinas on metal, any initial oxidation treatment establishes a barrier to further oxidation. This can thwart subsequent patination processes and result in weak or spotty patina on the surface.

For any patina on zinc, it must adhere to the surface. Some are thick and will fracture on bending, others are thin and will allow shaping. The initial patina must etch into the surface, essentially becoming part of the surface of zinc. This is done by an activation process on the surface you need to "wake it up" in order to get the reaction. After the surface is active it will seek out other compounds. The delicate nature of this period of time is short and you must neutralize the reaction once it begins to stabilize the patina and to relax the zinc.

For zinc, the patinas typically develop dark earth tones or white oxides. The darker tones tend to lighten on exposure as zinc hydroxide or zinc chloride forms on the surface. Figure 3.36 shows some additional patinas on zinc.

For the most part, these patinas are produced cold rather than with addition of heat like the copper patinas. Some of these take several days to fully develop. Like copper and preweathered steel, you want time for the zinc to diffuse out into the patina surface to make a metallic bond with the patina. Heating zinc to produce the patina needs to be kept at levels below 100°C. You do not want to dry out the surface nor damage the zinc.

One interesting surface that can be developed utilizes heat in a controlled application. Figure 3.37 is an image of a series of zinc plates that has been partially melted until the metal flows. It is not a patina but the surface darkens as carbon dioxide is absorbed.

FIGURE 3.36   Additional patinas on zinc.

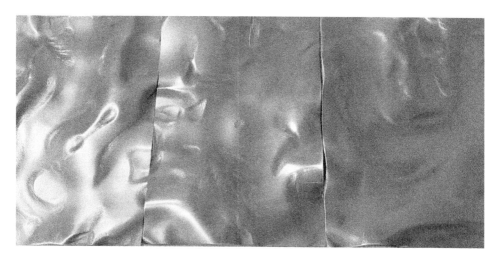

FIGURE 3.37 Melted zinc plate.

The caution is to avoid causing the zinc to boil and transition into vapor. Zinc is difficult to work this way because once it begins to soften, it falls away. There is a fine line between the time the surface liquifies and flows and the time it sluffs away. This surface is in development and begins with a thicker sheet of metal.

ZINC IRIDESCENT PATINA

Zinc, similar to other metals, can develop clear interference coatings of purples and golden bronzes with shades of browns and blacks. They are somewhat uneven and are produced by the reactions of the zinc surface to create a film of molybdenum oxide and zinc oxide. The color produced and the coating, see Figure 3.38, is durable and adherent. The coating can be further burnished and protected with wax. The zinc surface needs to be clean and free of oxides, then immersed in a molybdenum solution for a few seconds. The result is an iridescent coating on the zinc. The color produced is highly variable. Purple and red hues, coupled with bronze to backs, can be obtained.

GALVANIZED ZINC SURFACES

One of the most economical surfaces is the zinc-coated steel surface known as *galvanize*. Galvanized steel offers both economy and durability in most architectural exposures. The use as a protective coating for steel is the most common end use of the metal zinc. More than half of all the zinc processed each year goes to the protection of steel by immersion into molten baths of zinc or zinc-based alloys.

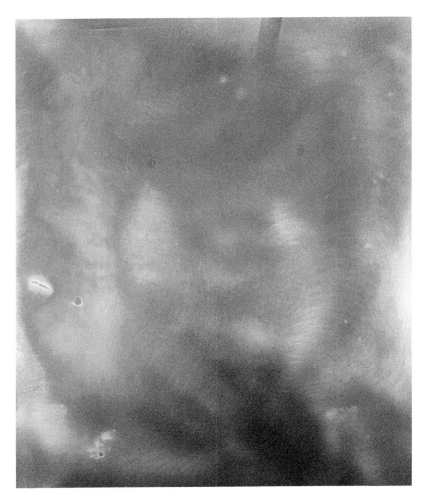

FIGURE 3.38   Iridescent patina on zinc.

Galvanizing steel or iron has been in use for nearly two centuries. First developed in Europe around 1837, coating steel or iron with zinc offers both an effective barrier and a sacrificial material to protect steel against corrosion. It is the barrier that offers the most protection to the steel, but the ability of extending protection across a breach in the coating or to uncoated edges is one of the unique attributes of zinc in its relationship to steel.

The hot-dipped galvanizing process involves taking steel sheet, plate, structural forms and even fabricated forms and immersing them in a molten bath of zinc. When this occurs, a very strong metallurgical bond is created between the two metals. At the interface of zinc to the steel substrate, there is a diffusion of atoms of zinc into the steel and iron atoms into the zinc. At this interface of the two metals, both alloying and iron–zinc compounds form. There is a layering affect that occurs as

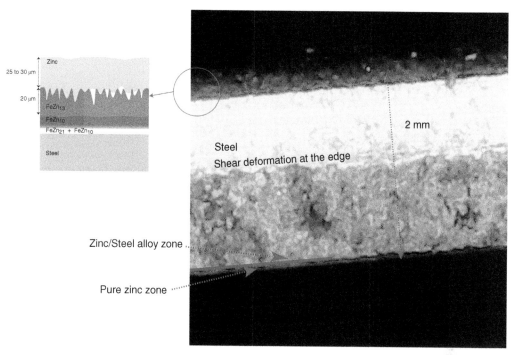

FIGURE 3.39 Edge of galvanized steel.

one metal combines with the other at different rates and complexities. There are four distinct layers of combinations of metals that form on the steel. Figure 3.39 show a close up view of what occurs.

The outer layer of the galvanized steel is high-purity zinc. It is the last layer developed as the steel is pulled out of the kettle of molten metal. This layer is tough and flexible and is 25–30 μm thick. The next layer below this pure zinc is a compound of $FeZn_{13}$. This is where the zinc and iron begin to diffuse into one another. This layer is about 20 μm in thickness, as the zinc iron diffuses into one another. This layer is hard and brittle compared to the outer zinc layer. Below this layer is a thin zone of $FeZn_{10}$, and below this is a mixture of $FeZn_{10}$ and $FeZn_{21}$. This layer is strongly bonded to the steel and to the alloy layer above. There are certain conditions where the galvanized can "peel" from the surface, but these are rare in art and architecture. Peeling can occur when there is contamination in the galvanizing tank and this contamination interferes with the bonding of the zinc layers. Once the process is performed correctly, the zinc layer is a near-integral part of the steel.

It is this multilayered jacket of zinc and zinc compounds that furnishes steel with a formidable protective barrier. The barrier is not porous, preventing water, air or other substances from reaching the steel. As long as this layered barrier is intact, the underlying steel will be protected.

From a designer's viewpoint it is the crystalline surface appearance, commonly referred to as a spangle that is most intriguing. This spangled appearance is not always apparent because of the manufacturing processes in use today. Most galvanized steel is destined for industries where appearance of the zinc surface is not of great concern. In fact, much of the galvanized sheet steel is produced for the automotive and appliance industries that further coat the surface with a decorative paint film. The spangle creates a problem as the rougher texture can telegraph through thin paint coatings. Therefore, the galvanized surface undergoes a treatment that eliminates the spangle altogether.

The heating and ventilation industry use a significant amount of galvanized steel sheet, as well. Here the spangle is irrelevant but contaminants in the zinc bath are a concern. Lead-free or nearly lead-free zinc coatings are required by industry. Much of the galvanized steel used to create ducting is applied by electroplating.

Electroplating methods differ from the hot-dipped galvanized process. The electroplated zinc are thinner coatings and the spangle is much smaller and tighter. This produces a high-purity zinc coating on the surface. No metallurgical bond is achieved but an even layer of fine crystals of zinc are plated to the steel. The grain is more cellular than what is achieved in the hot-dip method, which can be what the design calls for. Just understand, the coating is not as corrosion resistant as a hot-dipped galvanized surface.

For many art and architectural projects where unpainted galvanized is the finish choice, the spangle adds a unique feature to the surface and is a desirable attribute of the process. The crystalline spangle that forms on the surface is, on close examination, a six-pointed, flat manifestation of the zinc hexagonal, crystal lattice. The spangle as the surface crystal is often called, takes the form of a six-fold star pattern. The development of the spangle occurs when the molten zinc is cooled below the melting point of zinc, 419°C (787°F). The randomly arranged atoms begin to slow down and solidify into an orderly arranged crystal lattice. Tiny areas on the surface called grains of molten zinc attached to the steel and the cooling of the surface begins. The grains grow in a process called *nucleation* where the atoms begin to arrange into crystals at the initial state. If the cooling occurs rapidly there will be numerous small crystals formed. Slow cooling and the crystals become larger. The crystal arms of the star are referred to as dendrites. The rate of the dendrite grown is inversely proportional to the rate of nucleation. The dendrites grow out from the center of the star form as the zinc cools.

The dendrites need something to trigger the location of the center and cause the growth expansion out from this point. The presence of other elements such as lead, antimony or minor surface abnormalities on the steel will initiate the point of growth.

Galvanized coatings used in the early part of the twentieth century had around 1% lead while today there is 0.15% to as low as 0.05% lead. Lead effectively reduces the number of nucleation sites by decreasing the energy of the interface where solid steel meets liquid zinc in the galvanizing bath. Figure 3.40 shows the spangle on different sheets of steel and a microscopic view of the crystal.

As the galvanized surface is exposed to the atmosphere, the zinc on the surface weathers. The shininess of newly galvanized steel disappears. The spangle retains its crystalline nature and often the distinctive boundaries between the crystals become more apparent and appear as overlapping light and dark surface features. The left two images on Figure 3.41 shows a galvanized steel surface that has been exposed for several years. The right image is of a reflective, newly hot-dipped galvanized surface.

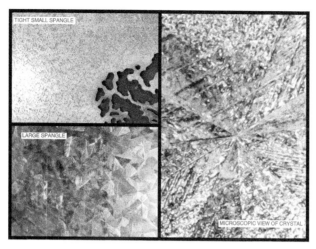

FIGURE 3.40 Spangle on steel.

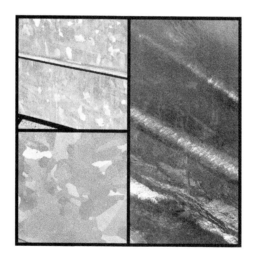

FIGURE 3.41 Aged galvanized steel.

On exposure, the zinc surface will convert to zinc oxide rather quickly. As water is present, the surface changes to zinc hydroxide. Further exposure of the zinc surface converts to zinc carbonate. At this point, the layer of zinc carbonate offers very good protection against corrosion for both the zinc and the coated steel.

The crystalline spangle that forms on the hot-dipped steel surface can vary significantly, depending on the chemistry in the zinc bath and the cooling rate. The zinc bath for most galvanizing operations contains 98% zinc. Often, other elements with different rates of solidification are in the bath.

As the part is brought out of the molten zinc, a layer of zinc adheres to the steel. Below this layer are the other zones with intermediate compounds of zinc–iron. This outer layer of pure zinc immediately begins to cool and the crystal initiates around the surface impurity and forms the spangle.

The longer the steel part is immersed, the thicker the coating will become. It starts quickly, but the process of building up zinc slows way down. The temperature of the bath is held at 455°C (850°F). The melting point of zinc is around 420°C, and zinc will boil around 907°C. For obvious reasons, one does not want to boil the molten metal. When the steel parts go into the bath of molten zinc, there is a vigorous reaction at first as the steel surface and zinc are developing a tight interface condition. This vigorous activity slows down after a few minutes as an equilibrium begins to form and the steel and zinc reach the same temperature.

For hot-dipped galvanized, the coating thickness, appearance, and adhesion are defined in ASTM A123 Standard Specification for Zinc Coatings on Iron and Steel and further in CSA G 164-18 for hot-dipped galvanizing steel shapes of irregular cross-section.

Galvanizing steel is an excellent option over painting. The coating of zinc is very durable and when the hot dipping process is performed with the visual surface in mind the results can be effective. When it is determined to have the zinc coating on steel as the finish surface it is wise to establish acceptable guidelines for the fabrication, handling and galvanizing process.

In general, when hot-dip galvanizing is used, the steel part should achieve the following:

- There should be no uncoated areas where zinc is not present.
- There should be no blisters where zinc has created a bubble on the surface.
- There should be no powdery substances on the surface from the flux or dross.
- There should be no inclusions on the surface due to dirt in tank.
- There should be no strings or runners from solidifying zinc drips.

Most galvanizing facilities are production facilities that process industrial shapes. Their goal is to produce the right thickness coating across the steel form as efficiently as possible. Finish surface appearance is not high on their quality assurance process. It is important that when specifying the use of hot-dipped galvanized steel for art or architectural forms, the surface quality requirements are understood.

Galvanized steel is so ubiquitous that it is often seen as a commodity and handled rougher than say copper or zinc sheeting. Scratches, mars, and moisture stains are overlooked and considered part of the metal.

If you do not set out instructions beyond "galvanized steel," the result from an aesthetic standpoint can be a disaster. Your parts will have the specified thickness, but they may have boot prints, scratches, drag marks, clumps of thickened zinc, or other surface conditions caused by poor handling.

Galvanized parts can be bulky and oddly shaped. When they come out of the zinc kettle, they are hanging from a gantry and they are very hot. The galvanized processor unloads them to an area where they are allowed to cool. The parts may be piled on one another as they cool. Boot prints may be imparted to the parts as the workers walk on the lower sections to get to the ones on top. The surface is simply not thought of as a special surface. Guidelines need to be established early on in the process.

Additionally, if there are large flat regions in the fabrication, they may come back warped from the heat of galvanizing. This can be prevented, but it requires an understanding from the galvanizing facility. You may need to add stiffeners or use a thicker cross section.

Design of parts to be galvanized is critical. Enclosed parts will float on the zinc. They need holes to allow the fabrication to sink into the molten metal and then again for the metal to flow out.

All of these challenges can be avoided by working with the galvanizing facility and making them aware of the end product use. The cost may be higher because of the handling constraints but the results will be improved.

> **Important Steps Critical for Successful Galvanizing**
>
> - Coordinate the part design with the galvanizing facility.
> - Eliminate closed geometry. Place holes to drain zinc out and for air to escape.
> - Discuss how the parts are to be handled when removed from the kettle.
> - For thin flat areas, discuss need for the pieces to cool slowly.
> - Avoid fingerprinting of the new zinc surface.

GALVANIZED STEEL STRUCTURAL SHAPES

Galvanizing structural shapes used to hold art and architectural forms are common throughout the world. These structures are pregalvanized and bolted together rather than field welded. Field weld damages the zinc protective layer and requires field fixes that do not equal the protective behavior of the hot-dipping process and that leave a visual blemish on the surface. When exposing galvanized structures, see Figure 3.42, care should be taken not to scratch or mar the surface. Workers should use gloves to avoid fingerprinting and clean the surface down after installation. A properly installed hot-dipped galvanized structure will perform for decades.

There is a tendency to use zinc rich primers, zinc solders, or zinc metallizing to repair damaged galvanized structural shapes. These work well in providing the cathodic protection and barrier, but when the work is exposed, these will age and weather differently, causing a visual blemish. For architectural exposed galvanized steel, any damage repair should be minimal and restricted to a very small area. The tendency often is to make a wide swath over the damage. This will lead to visually offensive splotches on the steel. There are numerous standards and standard practices that cover

FIGURE 3.42   Chrysalis structure made of galvanized curved tubing.

the repair of damaged galvanized steel but none of these take in the appearance concerns the design was trying to achieve in the first place. They may initially appear, but after a few months or years of exposure, the repairs stand out like spots on a Dalmatian. Touching up scratches on galvanized structures should be approached with this in mind.

DARKENING GALVANIZED STEEL

Darkening the galvanized surface is possible. Darkening galvanized is similar to darkening rolled zinc sheet or plate. A copper sulfate treatment will result in darkening the outer layer of newly galvanized steel, where the zinc outer layer is clean and free of oxide. Due to the way the surface reacts however, it is difficult to get a consistent tone. The edges of galvanized plates may be different. It is either very mottled, see Figure 3.43, or it is dark and somewhat dusty. It is recommended to coat the surface with a clear coating after darkening. This will prevent the white zinc hydroxide from turning a blackened surface to a black and white mottled surface. The right image on Figure 3.43 has one half coated while the other half is left to react with the environment and form a white crust of zinc hydroxide.

The darkening of galvanized steel must begin with a thick galvanized coating. Chemical darkening should only target the outer zinc layer. Therefore, the thicker the zinc layer the better the reaction and the more protection remains to the underlying steel. Figure 3.44 is the Morimoto Restaurant on 10th Avenue, New York, designed by Tadeo Ando. The façade has blackened galvanized steel panels at street level. After nearly a decade, there is some whitish spots developing from zinc hydroxide and possibly zinc chloride deposits. The zinc chloride will form when de-icing salts are allowed to reach the surface and remain long enough for moisture to create an electrolyte.

FIGURE 3.43 Darkening galvanized steel.

FIGURE 3.44 Darkened façade of the Morimoto's Restaurant in New York. Designed by Tadeo Ando.

ZINC PHOSPHATE COATINGS ON GALVANIZED STEEL

Coating galvanized steel with a zinc phosphate will produce a granular, even dark gray color that has excellent corrosion resistance. It has been in use for nearly a century and is very similar to what is created on some of the preweathered zinc surfaces. The finish has been used as a conversion coating to receive paint and improve wear resistance. The surface must be very clean to begin with, and fully degreased in detergent or industrial degreaser. Next, the coating is applied hot either by spraying the surface with a dilute solution or dipping. The solution is a proprietary phosphate solution that creates a noncrystalline, layer of zinc phosphate on the surface. Similar to a Parkerizing solution, these treatments create darkened conversion films on the surface of zinc or steel.

ZINC FABRIC

A method of coating zinc onto mesh to create a fabric-like surface has been developed. It is discussed further in Chapter 6, but the process involves shaping the mesh and immersing in a molten bath of zinc. Figures 3.45 and 3.46 shows an example of zinc 'fabric'.

The process has its challenges, but large sections have been created by shaping the mesh material, cleaning and degreasing the surface and immersing it in molten metal. The zinc coating will age and oxidize over time and exposure. The surface is very hard and robust. It can flex to a point before the zinc cracks.

Figure 3.46 shows larger test sections. It was necessary to understand how to coat the flexible material in molten metal without taking the shape out. As the material is removed from the bath, the weight of the zinc as it solidifies removes creases from the surface.

The beauty is the way the zinc color and reflective nature of the surface give the feeling of a soft, chiffon appearance but to the touch this is a "frozen," heavy metal material. It has interesting design capabilities as discussed further in Chapter 6, Fabrication.

FIGURE 3.45 Zinc "fabric."

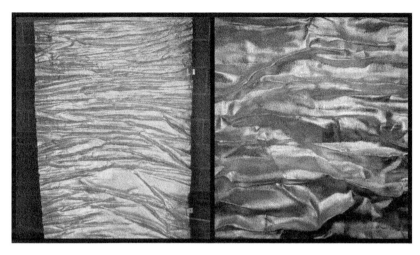

FIGURE 3.46   Test panels of zinc "fabric."

OTHER METHODS OF APPLYING ZINC TO STEEL

There are several other methods of applying zinc to steel. Zinc can be applied to steel by electroplating, sherardizing, thermal spray, and a process of tumbling. These processes, however, do not afford the same degree of protection to the steel and aesthetically they are not that attractive.

Electroplating lacks the zinc–steel metallurgical bond created by diffusion of each metal into the other at the interface. Electroplating imparts a thin layer of pure zinc onto the surface. The layer is even and bright but acts as a barrier coating with some level of sacrificial protection afforded by the thin zinc layer.

Sherardizing applies zinc to smaller steel parts by tumbling the parts in a heated chamber containing zinc powder. The powder bonds to the surface of the steel in a thin protective layer. This process gives excellent corrosion resistance, but the parts are darker in color and the process is limited to small sizes.

Thermal spray is a process whereby zinc wire is vaporized and blasted onto the surface of steel by means of a plasma gun. The surface produced is thick and porous. The appearance is grainy, but it can be honed down to create a smoother surface. For most applications due to the porosity, it will require an additional coating. On large areas, shaping can occur due to preprocesses involving the need to abrade the surface first. It is not used on art and architecture other than as a filler material to aid in corrosion resistance and leveling across seams.

Tumbling small parts in a rotating drum mechanically bonds zinc powder to the surface. Little corrosion resistance is achieved, and the parts are limited in size. This is used mainly on bolts and small fittings.

ZINC ANODIZING

Zinc can be anodized, but the process has never had commercial appeal. It is very expensive to operate and involves significantly higher voltage and a chemical process that involves ammonium hydroxide, chromic acid, phosphoric acid, and ammonium fluoride. The process improves the corrosion resistance by developing a hard, porous outer layer composed of oxides, phosphates, and chromates. The layer develops similar to aluminum by inducing a current – in the case of zinc, an alternating current rather than the normal direct current used to anodize aluminum.

The anodized coating developed is very adherent. Colors are limited to gray, green, brown, and charcoal. Zinc anodizing could be performed on rolled zinc, cast zinc, even galvanized coatings of zinc. The surface has been described as coarse, fritted surface of fused particles. At the time of this writing there is no known architectural source of anodized zinc.

CHAPTER 4

Expectations

....the remnant of conversations had over a glass of wine

Description of stains on a zinc bar top

INTRODUCTION

Zinc has been used as an architectural material from the time of Napoleon, yet many designers still fail to understand the nuances presented by this metal. We experience zinc as a surface material at every turn, we just do not recognize it as the metal zinc. Galvanized coatings are ubiquitous to our everyday world, often hidden below paint; zinc coatings in this form are extremely common.

This metal jacket of zinc we use so often to protect our steel gains mechanical strength from the metal being clad. It merely needs to be attached to the steel surface and be flexible, bending when the steel bends, expanding when the steel expands. This is taken for granted, but in reality, the zinc has made a tight, metallurgical bond with the underlying steel. So tight is this marriage of the two metals that at the interface there is a blend of steel and zinc, one partially diffused into another. This solid jacket of zinc can be applied to thick or thin, flat or fabricated steel. Figure 4.1 shows a solid steel sculpture created by flame carving the surface of a thick plate of steel by the artist Reilly Hoffman, then hot dipped in zinc to produce a protective, impervious coating. Unlike plating that creates a chemical bond or painting that adheres to a surface by mechanical interlocking into a surface, the zinc in galvanized coatings diffuses into the steel and the steel into the zinc making an interface zone of iron rich zinc.

In the context of art and architecture today, zinc is always in an alloy form. The alloy widely used in the cladding and roofing surfaces is the zinc–copper–titanium alloy. Zinc is alloyed with very small amounts of copper and titanium there are actually several versions of this alloy with very small differences. We will consider it as a single alloy for now since the properties are so similar.

FIGURE 4.1   Carved steel sculpture by Reilly Hoffman. Hot-dipped galvanized coating. Courtesy of R+K Design.

The use of this alloy is only a relatively recent development going back to the late 1960s and early 1970s. Prior to that time, alloys of high-purity zinc with traces of lead and other elements were used. These had limitations because of a mechanical deficiency, commonly referred to as creep, required thicker sheets to be used to offset the lack of strength. Creep is a condition where the zinc would undergo plastic deformation over time under relatively light loads at room temperatures. Another name for the phenomena is cold flow, since the metal moves at room temperature. This movement is not like flowing molten metal but a subtle change in geometry that works over time. Kind of like the slow movement of a glacier.

On a percentage by weight, only 0.20% of copper along with 0.12% titanium is all that is required to alter the mechanical and performance properties of zinc sufficient enough to make the metal a viable, thin cladding material; see Figure 4.2. These miniscule amounts go into solution and

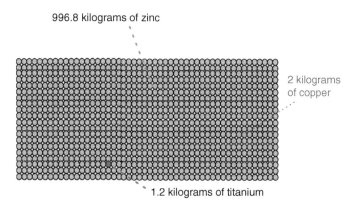

FIGURE 4.2 The percentage of copper and titanium added to zinc to transform 1000 kilograms.

disperse throughout the zinc lattice. This provides sufficient strength character to counter enough creep to make the metal comparable in mechanical behavior to copper and some of the aluminum alloys used in architecture.

These small, almost trace amounts of copper and titanium are all that zinc needs to change it. The maximum solid solubility of copper into zinc is 2.7%. Small amounts of copper are all that is needed to increase hardness and tensile strength and reduce creep. Titanium has even less solubility in zinc. At around 0.19% an intermetallic compound of $TiZn_{15}$ forms and is widely dispersed in the molten solution. When this compound forms in the zinc alloy, it restricts growth of the grain, causing the compound to solidify with markedly smaller grains. This compound also reduces creep. Adding both copper and titanium in these very small amounts to the molten zinc and creating the zinc–copper–titanium alloy gives it the necessary mechanical strength characteristic needed so that zinc could be considered for art and architectural cladding.

NATURAL FINISH ON THIN SHEET MATERIAL

The natural finish of zinc is a reflective, blue-gray color. In art and architecture, it is not a common finish used. The reflective nature of the mill, natural finish surface, on thin sheet and coil material is called "bright-rolled" finish. This reflective surface shows internal stresses, an undesirable manifestation commonly referred to as "oil canning." Oil canning can be reduced by embossing the surface to break up the reflectivity and adding a level of stiffness.

Texture rolling, a coining of the surface where one side has a light texture induced, is another way of reducing the appearance of oil-canning stresses. Essentially this converts a reflective specular surface into a more diffuse, light scattering surface.

On thin sheet, the zinc is rolled through a series of cold-reducing rolls. These rolls are pristine, polished rolls that impart a smooth texture to the surface as the sheet is processed. This gives the surface an initial glossy appearance. Any out-of-plane distortion will be enhanced.

The finish will weather; however, at first it will tarnish with a layer of zinc hydroxide forming rapidly, within hours of installation. This will not be evenly dispersed over the surface as it develops. There will be areas where the oxide will be more significant than others. Fingerprinting will also be apparent and if allowed to remain on the surface will etch the surface. The surface will darken more rapidly where moisture collects and will spot as differential drying occurs. This all happens quickly, in the first few days or weeks of exposure depending on the moisture and pollutants that are present.

As the natural surface weathers, how the surface will change will be highly dependent on the time the surface is wet and the pollutants in the atmosphere. The time of wetness is dependent on the relative humidity of the location the surface will be exposed in, the slope or how well the surface drains and if the surface is exposed to the drying effects of the sun.

Airborne pollutants or human induced pollutants such as deicing salts and glass cleaners, if they are allowed to remain on the surface, they will alter the appearance of the oxide and could induce streaking and spotting of the surface. Usually this will take the form of a dark spot. Figure 4.3 shows a dark spot on a galvanized steel surface. The spot is an alkaline cleaner that was left on the surface and has slightly etched the metal.

If the surface is installed clean, free of oils and fingerprints, and allowed to weather naturally, it will eventually develop the zinc hydroxide and then the zinc carbonate. The appearance will darken to a blue gray.

On the thin-rolled-sheet zinc, the natural unweathered finish is rarely used. Small castings that are shrouded in plastic housings, small engine carburetors or perhaps keys are made and left with the natural zinc finish.

Galvanized steel is the common use of natural zinc. Wrought forms of thin sheet zinc used in art and architecture are provided in the preweathered surface finish. The preweathered surface is a nonreactive, corrosion resistant surface.

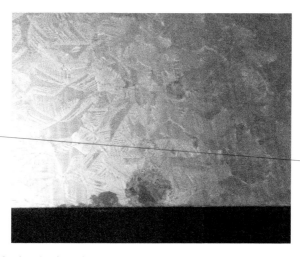

FIGURE 4.3 Dark spot of galvanized steel.

FIGURE 4.4   Zinc countertop. Image courtesy of Gary Davis.

Countertops and tabletops, once common back in the 1920s, are now experiencing a revival as artistic "aged" zinc. Made from zinc sheet that was in the natural finish, often of an alloy that was softer and more malleable, these bar tops would be glued down or attached to a wood substrate. Bar tops made of zinc once were so common the French used to refer to their pubs as *le zinc*. Bar tops made of zinc sheet with elaborate edge work or cast into thin plates were common fare. They stain easily from wine and beer spills, but they are relatively easy to clean and rubbing down the surface gives an attractive aged look. One manufacture describes the stains as, "the remnant of conversations had over a glass of wine." Figure 4.4 shows an elegant, modern bar top in a residence.

NATURAL FINISH ON THICK PLATE MATERIAL

The natural finish provided on thicker plate material, material with a thickness greater than 4 mm, is a hot rolled, as cast surface. Figure 4.5 shows a typical, as cast surface of thickened zinc with the as fabricated finish. The finish is rougher, lacking the smoothness imparted by the cold-rolling process. The surface has a less-refined grain. This tough surface has a natural, rugged beauty to it. Because of the roughness, the surface will hold more airborne substances. Cleaning with a pressure washer is sufficient to remove the dirt and debris that will collect on the surface. Over time, the surface will

FIGURE 4.5 Natural zinc planters after several years of exposure.

darken but the contrasting light-colored streaks will remain as moisture induces the formation of the hydroxide.

The thick plate material has small surface inclusions not unlike a hot-rolled steel surface. The material is rugged in both appearance and mechanical behavior. It lacks stiffness that you might see in a similar thickness of steel plate and these particular planters have additional, internal stiffener angles to bolster the form and resist shaping from the load of the soil.

The thick zinc will last a very long time. There will be no corrosion staining or rusting along the edges. These are zinc throughout and not zinc coated. When the useful life as a planter box is over, they are fully recyclable.

NATURAL FINISH ON CAST SURFACE

The cast surface of zinc will weather in a similar fashion to the plate surface. The cast zinc surface is smooth, when cast in a ceramic or plaster mold or coarse, when sand cast. Sand cast surfaces can be abrasion blasted to brighten them, but on exposure they will oxidize and darken as the surface transitions to a zinc carbonate.

The initial surface resembles cast aluminum in color. It is not until you handle the castings that you realize the distinction of the heavier zinc from the aluminum. Figure 4.6 show custom castings of zinc.

Zinc is easily cast. The low melting point and the fluidity of the metal makes zinc ideal for casting. It is not often used today as a cast material. Most architectural and art foundries are familiar with working molten bronze and do not appreciate what can be achieved with zinc casting.

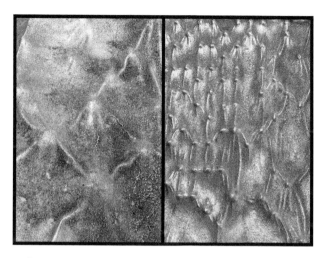

FIGURE 4.6 Cast zinc panels.

Zinc casting for other industries, in particular die casting is very common. Zinc is used to make numerous small detailed parts and assemblies for small engine parts, electrical appliances, chair legs, and untold everyday items.

Zinc casting of art statues and statuettes was common place in the late 1800s and early 1900s. The ones that have survived over the years have developed a fine texture and color of dark zinc carbonate on the surface from exposure to the atmosphere.

As long as they are occasionally cleaned the surface remains even. Usually these are provided polished or buffed to even the surface. Figure 4.7 shows a cast statue at the Regis College campus in Denver, Colorado. This statue was on display for decades in the main quadrangle of the college. It was restored and a new hand was cast and rewelded into place.

These cast zinc statues are thick walled hollow sculptures, usually greater than 5 mm thick. This thickness is needed to overcome creep, which over time leads to sagging and eventual failure.

Cast zinc is not as elastic as other metals used in art and architecture and can split or fracture if subjected to bending loads. The condition of creep can lead to rupture of cast zinc over time simply from the mass of the sculpture. Figure 4.8 is an example of several decades old zinc sculpture that have experienced creep. The sculpture on the left shows the crack forming in the foot while the one on the right shows a crack extending out on a copper-plated arm. The cracks resemble intergranular corrosion but on close inspection they are spidering around the zinc at these regions. Expansion from thermal movement probably enlarges these cracks. Zinc has one of the highest thermal expansion coefficients among architectural metals.

FIGURE 4.7 Cast statue at the Regis College Campus. Denver, Colorado.

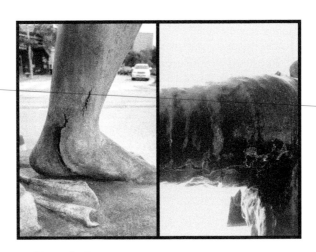

FIGURE 4.8 Zinc sculpture with cracks from creep.

PREWEATHERED FINISH

Once the strength characteristics of zinc were improved by the additions of copper and titanium, zinc sheet became a viable alternative to copper, lead-coated copper, terne-coated stainless, and other matte gray metals that were commonplace in the middle to latter part of the twentieth century.

Along with the improvements in alloying, modern methods of producing zinc flat sheet enhanced the production. Prior to the 1960s zinc sheet was produced by a process called *pack rolling*. Commercial pure zinc, 98.5% to 99.5%, was cast into a thick slab. Then rolled into thick plates. These plates were rough, lacked a fine grain, and were malleable. The plates would be sawed into smaller plates and assembled into packs. These packs would be heated and passed between rolls to reduce the thickness. Scale and oxides on the surface would prevent them from joining or welding together. Once the thickness was achieved, the individual sheets would be cross rolled to break up the grains and improve formability. The need for cross rolling limited the length and increased labor cost.

With these modern alloys containing copper and titanium, zinc could be continuously cast. Grain growth would be limited by the titanium-zinc compound, $TiZn_{15}$, as the metal was stretched and reduced in thickness in further cold rolling operations. Sheets could now be longer; coils of zinc could be produced in a continuous fashion.

The preweathered zinc surface is the most common form of zinc supplied to the art and architectural industry. All the major zinc mills involved with the manufacturer of zinc sheet provide the preweathered surface. Plate zinc and cast forms of zinc are preweathered by custom patination facilities and are not normally supplied by the mill with the treatments. The preweathering comes in one of two forms depending on the mill source. Some produce the zinc carbonate, refer to Figure 4.9.

FIGURE 4.9 Preweathered zinc with the zinc carbonate surface. 2 mm thickness.

114   Chapter 4   Expectations

FIGURE 4.10   Preweathered zinc with the zinc phosphate surface. 0.6 mm thickness.

This has a slightly green-gray hue. While others produce a zinc phosphate on the surface, Figure 4.10 shows an example of this surface tone.

The preweathered surface is a matte-gray to blue-gray appearance, sometimes with a hint of green-gray tones when the zinc carbonate is preweathered treatment. After preweathering, the ribbon of zinc is recoiled before being leveled and sheared into sheets. This can impart an initial sheen that gives the surface a level of glossiness, but soon after installation this glossy sheen disappears from the surface.

In normal atmospheric exposures, the preweathering creates a very stable outer layer that is highly resistant to change. They will age, but the aging is very slow and near imperceivable unless substances are concentrated on the surfaces and combine with the preweathered compounds.

FIGURE 4.11 Seaside roof.

For example, if the preweathered surface is used in a marine environment, zinc chloride compounds can develop. Figure 4.11 is a preweathered zinc roof installed near the sea in a humid environment. This type of exposure will rapidly form zinc hydroxide, which will mix with the chlorides that are prevalent in these environments and form a hydrated compound of a mineral form of zinc composed of zinc hydroxychloride, an insoluble white substance.

This will collect on the exposed surfaces and as more rains or condensation form this will redeposit along edges and drip zones. Figure 4.12 is the edge overhang of this same structure.

FIGURE 4.12 Deposits on the underside.

The sheltered overhang is showing the zinc hydroxychloride deposits left behind as condensation dries out each day. These are not damaging, only cosmetic. These are not easy to remove, nor is it practical. To prevent this a drip edge could have been established closer to the leading edge. A rib line running the length would be sufficient to break the capillary attraction and limited the amount of streaking.

Zinc exposed to marine environments performs well. Corrosion resistance in these environments is achieved by the formation of these dense corrosion products of zinc hydroxy chloride and zinc hydroxide. Both have low solubility in water.

The corrosion products that grow on the zinc surface are often lighter in color and therefore are more visible as they form and deposit on the surface of the preweathered zinc. One requirement considered as good practice when working with thin zinc sheets is to be sure the back side is adequately vented. Note the vent strip running the length of the overhang shown on Figure 4.12. The reason for this is to eliminate back-side corrosion of the sheet. When vented, the condensation that may form on the back surface dries out more rapid as air is allowed to reach it. Back-side corrosion of the surface usually takes the form of zinc hydroxide, but other substances can develop. If the area behind the zinc is enclosed or restricted, the pH level can drop depending on the influence of other materials that can be outgassing or changing over time. If this occurs there can be continual dissolution of the zinc and back-side corrosion can lead to premature perforation of the sheet. Sometimes you can see the decay manifest and leach out around the seams. Figure 4.13 shows white corrosion products coming from the seams in lapped, interlocked zinc sheet used on a vertical surface.

This is most likely coming from the reverse side oxidation. Condensation will form on the front and back side of the metal sheet. On the sloped portion, the condensation runs down the back side and collects in the seam. There also may be water entering above on this section of the roof. The seams of these thin sheets are folded back into a 180° hem. This hem holds moisture until it

FIGURE 4.13 Zinc corrosion products leaching out from the back side of zinc sheet.

can drain out the ends of the hem. While the moisture sets in the hem, the zinc oxide forms and some of it washes out onto the zinc below.

On preweathered zinc, dark streaks can sometimes be visible in urban environments. Refer to Figure 4.14. These streaks are typically one of three substances, steel items corroding above the zinc, zinc sulfide compounds, or general soot collecting on the surface. Steel set a distance above the zinc does not benefit from proximity to the zinc and the sacrificial quality of the metal. The steel can rust, and oxide particles will stain the zinc. Zinc sulfide, on the other hand, develops in urban environments containing combustion particles. Zinc sulfide in pure form is a white crystalline substances. However, impurities and contaminants will darken the appearance. The mineral sphalerite, for example, is dark grey to black. It is zinc sulfide with iron contamination. Zinc sulfide is water soluble and will dissolve from the surface of zinc exposing more zinc to continue to corrode.

Streaks can also be composed of other materials decaying and washing onto the surface of the zinc. Zinc has a tendency to combine with other metals and non-metals when the circumstances present themselves. The presence of moisture and a slightly acidic or alkali pH can create stains on the surface as these conditions can create decay in other substances near the zinc.

Preweathered sheet used in art and architecture should be well-drained surfaces. Water should not collect on the surface for any length of time. Much of this has to do with contaminants in the water that change the pH or establish corrosion cells on the zinc surface. Preweathered zinc gutters are in regular use on high-quality residential construction. In Europe, zinc guttering is commonplace on commercial buildings and residential buildings. Reference Figure 4.15. Zinc gutters are well constructed and add an elegance that defines the building edge. The guttering and downpipes are made from zinc sheet in thicknesses of 0.7 mm and 0.8 mm (0.027 inches and 0.031 inches). Zinc guttering is preweathered on both surfaces and rolled in half round sections. Downpipes, connectors, and other accessories are made from the same metal. All sides of the gutter are exposed to

FIGURE 4.14 Dark streaks on weathered zinc surface.

FIGURE 4.15 Zinc guttering.

the environment and gutters are designed to drain. From a standpoint of concentrating polluting substances, the gutter is usually a place of concern, but these elegant forms have a proven lifespan. If installed correctly, they can last several decades with little maintenance.

Thin roofing and cladding material made from zinc is preweathered on both sides. However, the concealed, hidden side that is placed up against a substrate should be designed to allow ventilation. Sometimes referred to as a breathable surface, there are several membranes on the market used to separate the zinc from the rigid wood or other solid surface. Figure 4.16 shows a thin, 0.8-mm-thick

FIGURE 4.16 Thin zinc sheets with a Hunter Patina over a preweathered surface. Green layer is the breathable membrane. Continuous cleat is needed for strength.

zinc shingle installed on one such substrate. Note the continuous stainless-steel cleat as well. Thin zinc is weak, and the continuous stainless steel cleat adds stiffness to the edge as it holds the shingle in place.

PREWEATHERED WITH ADDED PIGMENTATION

There are several manufacturers of rolled zinc product that add color to the zinc preweathered surface. The colors are semitransparent and add a distinctive hue to the zinc. Figure 4.17 is an image of one such finish imparted to the zinc surface. This blue tone is very durable and resistant to ultraviolet. When it is applied to the ribbon of zinc, the color is even and consistent across the surface. Its exact makeup is proprietary, but it resists mild solvents and adds a level of corrosion protection to the surface by nature of the added surface polymer that carries the pigment. The clear coating that carries the pigment is only 10–12 μm in thickness. So, its ultimate durability has limitations. The weathered zinc substrate is an excellent, nonreactive base for the pigmented lacquer.

The pigments are advertised as mineral colors, which suggests the slight color hue is induced from minerals rather than dyes. The evenness and color tend to make the surface look as though it is painted aluminum. Up close, the characteristic grain is apparent, but a few meters away, the surface looks similar to a painted metal surface.

These are new finishes to the market and how they will perform over time is dependent on how these thin, transparent coatings will perform in various environments.

FIGURE 4.17   Blue pigment coating on zinc.

EXPECTATIONS – PREWEATHERED SURFACE

A properly fabricated and installed preweathered zinc surface should last well over 80 years even with the thin rolled zinc material in use today. The preweathered zinc surface is very passive, and exposures to mild and light industrial environments should only enhance the surface by the development of adherent corrosion products. The corrosion products, consisting of zinc oxide and hydroxide, zinc carbonates, and zinc hydroxy chloride aid in the ultimate protection of the base metal. Once they form, they adhere to the surface and slow the rate of corrosion down.

Zinc exposed to the open air where moisture condenses on the surface of the metal each day will form zinc carbonate. This is due to condensing moisture normally contains carbon dioxide, which over time reacts with the zinc to form the carbonate. The zinc carbonate outer layer is what provides zinc with the corrosion resistance.

FIGURE 4.18 Preweathered zinc surface as it appeared in 2000 and as it appears in 2020.

The zinc phosphate used to darken or preweather the zinc surface also provides corrosion protection. These are usually thicker coatings and their appearance is darker. These will over time react with moisture and build out a surface of zinc carbonate.

In urban environments laden with sulfide compounds is where the zinc surfaces face challenges. Today, in most of Europe and the United States, sulfur compounds are greatly reduced. Sulfur mainly comes from sulfur dioxide and is responsible for what is known as acid rain.

The source for sulfur dioxide and other sulfur compounds is combustion of fossil fuels, mainly coal fired power plants. In the United States, the sulfur dioxide levels are low in most regions. Same with Europe where stringent environmental controls are maintained. The region around the Ohio Valley in the United States are the highest concentrations of sulfur dioxide as have also increased in certain regions of south central Mexico.

The largest sulfur dioxide region on Earth today is in China. A large section of the north along the upper eastern industrial part of the country. Zinc would not perform well in these regions of the world due to the high sulfur presence. Figure 4.18 shows a zinc surface installed in an urban environment 20 years ago in the left image and as it appears today in the right image.

BLACKENED ZINC

Zinc can be blackened in several ways and is available in a matte, charcoal black color. The black produced is very even and consistent across the surface of each sheet of zinc. Black can also be applied to cast parts and even assembled shapes.

In sheet forms, blackening is performed during the preweathering process.

The blackened surface ages and lightens slightly as the surface is exposed to the atmosphere. Moisture that collects on the surface from condensation or water vapor in the ambient air will diffuse zinc ions through the darkened layer. These zinc ions will combine to form zinc oxide and zinc hydroxide on the surface. Zinc hydroxide is white in color, and these tiny white crystals will form on the dark surface. This will proceed very slowly but the net effect is a lightening of the dark surface. Refer to Figure 4.19.

The more challenging aspect with the black zinc surface is to be certain the white zinc hydroxide stain does not develop. This can happen when shavings of zinc are not collected after drilling holes. There have been occurrences where the drilled shavings were allowed to remain or fall behind the panel being attached. The shavings oxidized and streaked over the blackened surface creating unsightly white corrosion streaks over the black zinc.

FIGURE 4.19 Slight lightening of the zinc surface from minute formations of zinc hydroxide.

COLOR MATCHING

All the preweathering and darkening processes are essentially transition layers on the zinc surface. They develop by chemical reactions with the zinc surface to form the oxides, phosphates, sulphates and carbonates that alter the color. When these processes are undertaken on sheet forms, there is a directionality to them. Altering the direction of the sheet can impart a slight color variation. However, the process itself of generating the transition layer can also influence color matching between element of differing production runs. Figure 4.20 shows a preweathered surface, where the altering bias is purposely used to create a cacophony of pattern variation.

It should be noted that this color variation occurs only under certain lighting conditions while other light conditions, say cloudy versus bright sunlight, the same variations disappear or reverse, this characteristic is indicative of the tight grain of the metal and how light interacts at a micro-level.

Other conditions are more dramatic. These are directly related to the surface variations that are difficult to control in the continuous casting, rolling, and reducing operations. In particular, the color can vary in the thicker sheets, 1 mm and greater, when they undergo preweathering treatments. Over time, this will adjust somewhat but still can take several years before the color blends. Figure 4.21 is a surface of 1 mm panels that show variation from one process batch to the next. This can be difficult to identify until installed and viewed from a distance. This condition can also be pronounced when the sheets are flipped during fabrication.

Angle of view can also influence whether color matches or variations are visible. When this phenomena occurs, it is very difficult to spot until the metal surface is in place. This condition happens with other natural metals as well. The preweathering will often enhance this condition. Over time, the variation will diminish, but this will take years, depending on the environmental exposure.

Color Matching 123

FIGURE 4.20 Zahner Engineering Space. Image courtesy of Dan Gierer.

FIGURE 4.21 Color variation visible in bright sunlight.

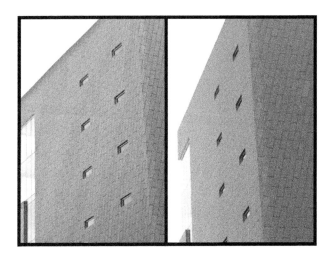

FIGURE 4.22   Variations in tone when viewed at different angles.

Figure 4.22 shows a surface where at one angle it matches well and at a slightly different angle the variations in batch runs is apparent.

CUSTOM PATINAS

There are several custom patinas made for zinc as presented in Chapter 3. These patinas take advantage of the zinc ion affinity for other metals and non-metals. In the case of the custom patinas, the effect of color variation from element to element is often desirable. This variation is to be expected and is not unlike the appearance of other natural substances such as wood and stone. They may come from the same family of tree or from the same quarry, but natural formation of diverse elements have a uniqueness that must be accepted for its inherent beauty. The patinas are composed of complex compounds that, when formed correctly on the zinc surface, are very durable and adherent.

Essentially these are minerals of zinc often layered to form an insoluble barrier to the environment. Since these are chemical reactions dependent on the nature of the zinc surface, temperature, humidity, and pH, the resulting patina is akin to the bark on a species of tree. No two are exactly alike. Figure 4.23 shows a surface clad in the Roano™ patina. This is a layering of several zinc compounds that forms over a short duration within the controlled confines of a factory. These patinas develop a passive surface on the zinc by combining with the free zinc ions at or near the surface. The compounds that form are adherent to the base metal and are insoluble. They do not come off in heavy rains and stain the adjoining surfaces.

Initially, there can be some minor dusting, but the first few wet–dry cycles usually eliminate this as the oxidation reactions continue and intermix with the base zinc.

It is important to remember that when these are produced on zinc surfaces, they should remain in a controlled environment where the relative humidity is high and the temperature is moderate.

Custom Patinas 125

FIGURE 4.23 Custom patina Roano™ Barnard College New York. Designed by SOM Architects. Image courtesy of Tex Jernigan and ARKO.

Air should be allowed to reach the surfaces. Moving the patinated surfaces too quickly will create adhesion issues. The surface must have time to react and form into the compounds.

The layering aspect happens as zinc diffuses through the first layer and reacts with the other ions in solution on the surface to form still other compounds. This is not like paint where the coatings are built up on the surface and must adhere to the surface below. This is a chemical reaction of various compounds with the zinc to form new compounds with different color tones.

As noted previously, there is no staining of adjacent surfaces from these patinas once they have bonded tightly to the underlying surface. Figure 4.24 shows a custom patina on zinc created for the

FIGURE 4.24 Max Brenner Chocolate Bar™ façade. Image courtesy of Tex Jernigan and ARKO.

Max Brenner™ Chocolate Bar façade. The surface finish demonstrates a durability and appearance without any staining of the concrete sidewalk below. The finish is called Baroque™.

Variations on the Roano™ patina can be generated to a point by adjusting chemistry and exposure times of the various layers. The surface can be darker by allowing more of the preweathered zinc to appear. Allowing more zinc hydroxide to develop can push the color tone to a slightly lighter version. Figure 4.25 shows a project designed by Diller Scofidio + Renfro. To create the surface appearance, the designer wanted several tones of Roano™ patina on the Stanford McMurtry Building for the Department of Art and Art History. This required close development with the design team and an understanding of what is possible to achieve.

Sometimes, these processes can be pushed too quickly for the proper development of the surface. Working closely with a design team and client to achieve a custom, mineralized surface on zinc can be daunting but rewarding when the end result is something new and special.

The layered patinas on zinc can appear lighter or darker depending on the angle of view and the lighting condition. This is due in part to the levels of oxides and other compounds on the surface, which have a light-scattering effect. The microscopic variations on the surface and the stochastic disbursement that occurs as they grow and develop can lead to significant light reflective changes as you move around the surface. The very low reflectivity and the way different compounds on the surface absorb and reflect different wavelengths of light lead to an interesting visual array on a macro level. Some of these compounds are semitransparent and allow light to pass through them and reflect the surface below. Figure 4.26 is a microscopic reading of a typical surface. These readings of height and surface spatial parameters are not unlike what you would get from a moderately smooth stone surface. The big difference is how some of the light hitting the surface is absorbed and re-release back to the viewer.

FIGURE 4.25 Stanford McMurtry Building, designed by Diller Scofidio and Renfro. Image courtesy of Tex Jernigan and ARKO.

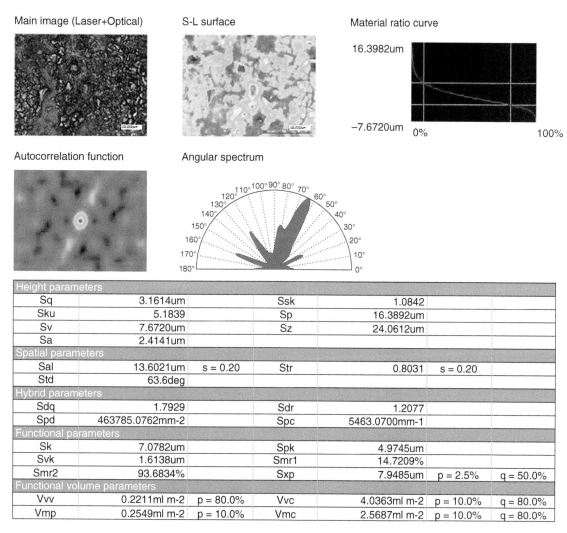

FIGURE 4.26 Microscopic images and special relationship of a Roano™ surface. The parameters listed relate to relative differences on the surface.

The microscopic roughness analysis shown in Figure 4.26 is an optical profile analysis. This type of surface analysis provides tremendous information about the surface of a material and how light interacts with the surface.

The various parameters show relative information about a sampling of the surface. The angular spectrum shows how different intensities of light reflect from the heavily patinated surface.

What this indicates is this surface will exhibit some angular differences as light reflects back to the viewer at different angles of view. These thick patinas capture light in the coarse micro landscape while reflecting other areas.

Figure 4.27 shows a surface of Roano™ zinc, installed in an interior museum space. The surface has been installed for 10 years with little to no cleaning. Granted, a museum space is going to have a very low relative humidity so changes to the surface should be minimal.

As you move around the surface, different angles show slightly different color tones.

These images were taken a few minutes apart on the same day at different angles of view. The camera has a more discerning eye that what the human eye perceives and can pick up lighting conditions the eye misses. Variations, particular on light-scattering surfaces viewed at acute angles can be accentuated more than the eye can pick up. This subtle change was visible as the angle of view changed only slightly.

What this shows is the depth of the finish, as indicated in the microscopic profile shown in Figure 4.26. This is the case with many patina finishes on metals, whether zinc, copper, or weathering steel. Unless the surface has been coated with a clear seal, light will reflect and scatter from the surface at different levels, leading to a tone shift as the angle changes.

On the exterior surface, exposed to the atmosphere, moisture and the accompanying components carried in the air, Roano™ patinas will change but only slightly. Figure 4.28 shows the Taubman Museum surface just after installation and after 10 years of exposure in the urban environment of Roanoke, Virginia. Roanoke is a very clean, light industrial city; however, a series of train tracks run parallel to the façade only a few meters away. So, exposure to the frequent diesel exhaust of these train engines would be expected to add to the pollution on the surface.

The surface had not been cleaned, other than occasional rains. There are whitish deposits of zinc hydroxide forming on some of the sheltered areas. But overall, the surface has hardly changed in 10 years of exposure.

FIGURE 4.27   Changes in color and appearance.

Custom Patinas 129

FIGURE 4.28 Roano™ surface on the Taubman Museum today and 10 years ago.

The white, dusty appearance forms along the folds of the flat Roano™ finished panels. This is a light deposit of zinc hydroxide that has developed over time where the folds hold dirt and moisture and redeposit it on the adjacent edge. It can be wiped or washed off. It makes the overall surface a bit lighter than the original installed appearance. You can also see it under the protected overhang areas and along the base below the aluminum vent. Reference Figure 4.29.

Expectations vary, depending on humidity and exposure. The patina applied to the zinc surface continues to adjust with this exposure as time passes. Figure 4.30 shows the Hunter™ patina and

FIGURE 4.29 Changes to the Roano™ patina after years of exposure.

FIGURE 4.30 Changes to the Hunter™ patina after several years of exposure. Top image is just after opening and lower image is 15 years later.

how the zinc surface has aged over 15 years of exposure along the banks of the Tennessee River in Chattanooga, Tennessee. The top image in Figure 4.30 shows the surface as it was first installed. The lower image is 15 years later. The lighting is different, but you can see regions, particularly where the surface is sheltered and back sloping, where a darker tone is apparent.

There is nothing damaging to the base metal in either of these images. It only changes in the oxide that show as light or darker tones, depending on the viewing angle, ambient light conditions, and what substances exist in the atmosphere that are reacting with the zinc and combining with the patina.

From the observations, it is apparent that sheltered regions are most susceptible to finish changes. Dark streaks or light streaks can develop as the zinc ages. These streaks are in part the result of other substances decaying and interacting with the zinc. Moisture in the form of condensation is remaining on the surface longer on sheltered areas and this same moisture comes from areas above where condensation has formed. Gravity pulls this moisture down over the surface, and as it travels it will pick up pollutants and other substances and redeposits them as it accumulates. As this moisture travels the surface, it concentrates the pollutants. Sheltered areas have limited rainfall exposure and do not benefit from the washing effects these rains provide. These conditions lead to sheltered regions having a greater inclination to being confronted with microclimates that can alter the appearance of the patina on the surface.

Maintenance plans for a building surface should identify the areas where sheltering is more pronounced and institute an occasional cleaning regimen. This would replace the lack of natural cleaning afforded by rains.

FIGURE 4.31 The venerable zinc roofs of Paris. Image curtesy of Gary Davis.

As time progresses, nature will take its course on these natural, mineral-like surfaces and provide the timeless beauty the metal zinc can express. The roofs of Paris reveal an elegant, aged appearance as time and exposure convert the surface of the zinc to a stable oxide. Dark and light tones reflect back to the viewer in a soft, but venerable way. Refer to Figure 4.31.

FLATNESS AND VISUAL DISTORTION

Whenever thin metal is used, minor surface distortions can be visible. These distortions, sometimes referred to by the colloquial term "oil canning," are caused by instability in the thin metal diaphragm that make up a low profile flat plate. A diaphragm is developed when there is a large expanse of unsupported, thin material. The instability occurs because the metal has uneven internal stresses that react and change as the surface undergoes thermal changes. A flat surface expanse can go concave or convex depending on the distribution of stresses. Localized areas may stretch or contract slightly different than adjoining areas. It is a balancing act as the surface seeks a level of equilibrium.

Chapter 6 discusses the thermal expansion characteristic of flat-rolled zinc and how zinc has an anisotropic behavior, that is expansion and contraction with the grain is different than against the grain. Figure 4.32 shows oil canning in the face of a thin blackened zinc panel supported along the edges. The edges are folded, so this locks the internal stresses into the flat face portion of the panel.

The newly installed zinc has a level of gloss that should diminish as the surface ages. Table 4.1 shows some typical gloss readings taken with a 60° glass meter. In the table, the most reflective

FIGURE 4.32   Black zinc panels displaying surface distortion.

TABLE 4.1   Relative Gloss of Various Zinc Finishes and Other Metals for Comparison

Metal and Surface	Relative Gloss
Natural finish zinc	200
Natural finish with fine texture	140
Zinc carbonate preweathered surface	33 to 40
Angel hair finish on preweathered surface	10
Black preweathered zinc	6
Phosphoric preweathered surface	1.5 to 3
Roano™ patina zinc	1.5
Baroque™ patina zinc	1.2
Stainless steel – glass bead finish	90
Stainless steel – angel hair finish	80
Anodized aluminum	8 to 30

surface is the new, natural finish zinc. These surfaces have an apparent grain, but they are very reflective due in part to the clean, cold rolls used to impart a thickness at the mill. All of these finishes will quickly lose gloss as they are exposed to the atmosphere. The preweathered zinc carbonate surface, when new has a reflective sheen. This too will weather further, and this should drop in gloss.

In the images, Figure 4.32 the surface appears glossy. This could be due to the clear coating the mill provided to give some initial protection. When it is first installed the light reflects off of the localized areas of concavity or convexity and show a slightly different appearance.

The distortions are most likely induced in the panel by a combination of factors. The thin zinc, in a black color, will absorb infrared radiation and get quite warm. As the zinc heats up it moves and expands. The flat portion of the zinc panel may have internal stress from the panel manufacturing where the edges were stretched at a different rate than the center. Additionally, handling of the panels can induce stress, but the most likely cause of the distortions are the fastener in the reveal. This pinned the panel and restricted expansion and contraction as the zinc attempted to move. Even if intermediate cleats were used at the supports, if they are not perfectly aligned, they will induce distortions in the face as the metal heats up and expands.

Figure 4.33 shows panels of similar thickness to the black zinc, 0.8 mm (0.032 in.) with a Roano™ finish. The finish helps conceal distortions that may be present in the face of these panels. The finish is very low in gloss, see Table 4.1 for comparisons of gloss. The eye is unable to distinguish surface variations if they were present, due to the variations in the Roano™ finish.

Figure 4.34 depicts how the panel could be assembled with a continuous cleated support. The continuous cleat along the edge allowed the metal to move without restricting the geometry as the zinc underwent temperature changes. The continuous cleat also provides strength to the edge allowing the zinc to move without the possibility of disengaging under negative loading induced by high winds.

Zinc is weak in the context of other metals used in art and architecture and therefore you need to consider augmenting the support and connections to accommodate this inherent weakness. Rolled-zinc sheet has different strength characteristics depending on grain direction. This anisotropic behavior has to be understood and acknowledged when designing with zinc sheet

FIGURE 4.33 Barnard College. Roano™ zinc patina on custom formed panel.

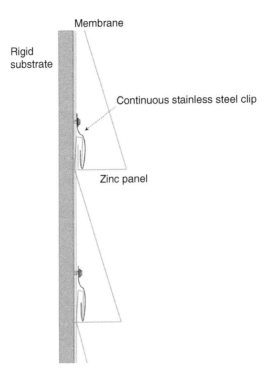

FIGURE 4.34 Section of panel with continuous stainless steel cleat.

otherwise potential issues can develop. Table 4.2 shows some of the anisotropic mechanical difference for zinc.

The Institute for Contemporary Art at Virginia Commonwealth University designed by Steven Holl Architects used 2-mm-thick zinc plates with a preweathered surface. The panels were 1 m wide and 2 m in length. These large flat panels have stiffeners on the back of the panel that act as part of the support system. This allows the panel to move as thermal changes occur. Figure 4.35 shows a corner of the building. The face fasteners hold the edge fold onto the panel. Every panel is allowed to float. The glossy preweathered surface shows no distortion because of the design of the attachment system and the thickness of the panel.

TABLE 4.2 Approximate Anisotropic Mechanical Properties of Rolled-Zinc Sheet

	Coefficient of Thermal Expansion	Tensile Strength
With the grain	$23.4/°C \times 10^{-6}$	165–207 MPa
Across the grain	$19.4/°C \times 10^{-6}$	207–262 MPa

FIGURE 4.35 2-mm-thick zinc panels for The Institute for Contemporary Art at Virginia Commonwealth University. Designed by Steven Holl Architects.

CREEP

All metals can experience the phenomena known as *creep*. This is because metals are made of crystals with metallic bonds between the crystals. Metals are malleable because they allow slipping to occur within the planes of these crystals. Creep is the elongation of a material when under a constant strain over time. An analogy can be made in the way a viscous substance moves and spreads out slowly under its own mass.

When metals are put under stress they can elongate. When the load is removed, they return to the original form. This is known as elasticity. When they do not return, they undergo plastic deformation. For creep to occur, the metal undergoes change in geometry over time. With the phenomena of creep, the plastic deformation happens when the stain is much lower than levels of stress that would lead to fracturing the material.

For zinc, in particularly the unalloyed forms, the crystal lattices that make up the metal slips over one another when under strain. Vacancies within the lattice allow dislocations in the crystal to move out of plane and fill the vacancies with other atoms. Generally, elevated temperatures are required for metals to flow and fill these vacancies. With zinc, the crystals will slip under room temperature at light loads.

Creep happens over time as the zinc surface undergoes strain. There are three distinct stages zinc can undergo when subject to creep. Figure 4.36 shows these stages as a relationship of time to strain.

The primary stage of creep occurs at a decreasing rate over time. It begins rapidly as the stain increases but slows to a steady-state condition as it moves into the secondary stage of creep where changes in strain are low. Elongation occurs slowly, but at a steady pace with time. Once in the tertiary stage, it can move rapidly to rupture.

This is why it was critical to develop a zinc alloy, the zinc–copper–titanium alloy that resists creep and adds strength and hardness. The small amounts of copper and titanium can fill the vacancies as the crystal planes slip, adding resistance to movement. Pure zinc, prior to this alloy development, was weak, stronger than lead or tin, but nowhere near the levels of copper and aluminum. Adding small quantities of copper and titanium causes the zinc alloy to undergo a metallurgical process known as *dispersion hardening*. While the zinc is in molten state, small amounts of copper and

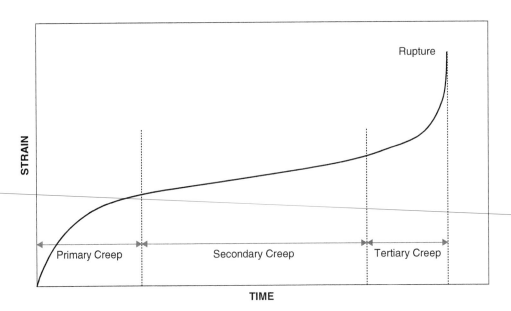

FIGURE 4.36 Stages of creep. Time versus strain.

titanium are added. Immediately, the titanium forms $TiZn_{15}$ and the copper atoms are absorbed into the zinc in what is referred to as *the eutectic mixture*. Strengthening occurs as these small, hard particles are dispersed throughout the crystalline lattice as the zinc is cooled. These small particles will restrict the slippage and movement of dislocations in the lattice and increase the strength of the zinc alloy. When the sheets are rolled out and heat treated the metal strength is greatly enhanced.

Copper added to zinc in very small amounts nearly triples the strength and increases creep resistance significantly. The drawback to the added copper is it increases the temperature where the zinc sheet is brittle and will fracture. When working with zinc, forming the metal at temperatures below 10°C (50°F) can crack the metal.

The small amounts of titanium improve the grain structure as the metal cools. Titanium creates a superfine grain structure and this, along with the copper can arrive at an ideal combination of ductility and strength.

An example of creep can occur with the relaxation of form. Figure 4.37 shows a perforated and corrugated screen made of preweathered zinc. These panels were custom perforated and custom corrugated at a diagonal. The diagonal corrugation crossed the grain at an angle, and thus the sectional strength achieved by corrugating was partially offset by differences in strength from crossing the grain at a diagonal. During shipping, the weight of the unsupported zinc panel and repetitive bouncing as the shipment moved on the truck for several hours, the zinc panel slightly flattened the corrugated form. Once installed vertically, the corrugated zinc screen has stood for over 25 years.

Creep is not the issue it once was when statues were created from zinc alloys lacking the copper and titanium. Creep is something to acknowledge, and it is good practice to restrict concentrations of strain in zinc panel assemblies.

FIGURE 4.37   Zoo corrugated panels.

GALVANIZED SURFACE

Designers and artists consider the galvanized surface for reasons of economy, durability, and its industrial appearance. Over the last several decades, the characteristic large spangle has been eliminated and replaced with an initial shiny surface with small, tight spangle hardly visible.

As described in Chapter 3, the spangle is a manifestation of the zinc hexagonal crystal. On close examination it is six-sided star-like shape that reflects light at slight angles from a diffuse, light scattering surface. Figure 4.38 shows hot dipped galvanized steel plates. The spangle is a medium size but still clearly visible.

It is interesting to note, the edge of these plates have, in some instances, a visible alignment of the crystals. This is because of the way they were removed from the molten zinc. The crystal growth initiated along the edge and cooled before the balance of the plate. This is similar to the crystal solidification that occurs in a casting. The outer edge cools first and the crystals extend inward from the walls of a casting.

There are generally two types of galvanizing processes that are considered in art and architecture for achieving the desired surface appearance. These are continuous galvanizing performed on coils of thin sheet and batch galvanizing performed on fabricated items, plates of steel, tubes, and structural forms.

The continuous process applies a layer of zinc on the edges and top and bottom surfaces of sheet steel while it is still in coil form. This process produces an even, consistent coating of zinc. It is generally a thinner coating and is the method used to supply galvanized sheet metal to the automotive, heating and ventilating, and roll-formed building panels.

FIGURE 4.38   Hot-dipped galvanized steel plates.

In the continuous galvanizing process, cold-rolled sheet steel in coil form is passed through a series of degreasing and pickling tanks before entering an annealing oven. As this moving strip of metal exits the oven, it is very clean and free of oxides. The strip enters through a protective atmosphere that activates the surface just before entering the tank of molten zinc. As the coated strip exits the molten zinc, an air knife removes excess zinc and begins the cooling process. The zinc-coated strip is leveled and recoiled.

The sheet material produced this way is destined for eventual forming and the layer of zinc needs to retain flexibility. For that reason, aluminum is often added to the molten zinc bath. Aluminum is added to initiate an iron aluminum layer and stop or slow down the diffusion of iron out into the zinc. This gives a high-purity zinc coating that lacks some of the more brittle zinc-iron interface layer and maintains a very flexible coating. The final surface is very even and consistent. The spangles are generally small and highly regular.

Batch galvanizing is performed by dipping steel articles, fabricated forms, plates and assemblies into a series of tanks. The parts can be large assemblies handled by overhead cranes that manipulate them into degreasing and pickling acid tanks to prepare the surface.

The parts are passed through a fluxing tank or layer of zinc chloride and ammonium chloride and then into the molten zinc bath. After a certain period of time, the parts are removed from the tank and set to cool. Figure 4.39 shows a series of tubes with welded anchors that were galvanized by the batch process.

The thickness of the zinc is variable across the surface when batch dipped in molten zinc, but much thicker than when continuously galvanized. Shapes, interior surfaces, welded assemblies can all be batch galvanized. Batch galvanized steel parts are normally not post formed or shaped.

FIGURE 4.39 Tubes with welded clips. Batch galvanized.

Therefore, aluminum is not typically added, and the zinc-iron interlayers develop. These are brittle and less flexible. Shaping can split and crack the galvanized coating at these layers.

In batch anodizing the surface may have glistening appearance with waves or bands that develop as the pieces are removed from the molten zinc bath. The surface can also possess a marbling appearance similar to the spangle of old but much less defined.

Both types of galvanized coatings will age similarly. The shininess of the new coating will quickly develop a darker, low reflective tone as a thin layer of zinc oxide develops. The zinc oxide and zinc hydroxide form on the surface within a matter of days. As exposure continues, this becomes the protective zinc carbonate. Figure 4.40 shows weathered continuous and batch galvanized surfaces after a couple of decades of exposure.

The upper image in Figure 4.40 is a corrugated sheet made from a coil of steel that underwent the continuous galvanized process. The spangle is no longer visible, and the surface has an even gray tone. As the zinc wears away, the steel will become exposed and can corrode.

The lower image is an assembly that was batch galvanized. You can see the welds used to fix the plates into the steel frame are galvanized as well. Thus, the railing system was assembled from steel bars and plates and hot dipped as an assembly.

Table 4.3 is a comparison of attributes of continuous galvanizing versus batch galvanizing. They both are production processes. The continuous process is rapid and set for high speed processing of coils whereas the batch process is established for piece processing of more irregular forms.

Defects are more prone to occur in batch galvanizing. The fact that it is often approached as a production process can lead to lack of artistic or aesthetic concern. Production processes want to move rapidly. Additionally, the uniqueness of the parts can lead to uncovering new challenges that are unwelcome. Figure 4.41 shows a hot-dipped batch galvanized steel frame for an architectural part. The zinc is roughened along the edge as if it is contaminated or not adhering to the steel.

FIGURE 4.40 Weathered galvanized steel.

Galvanized Surface

TABLE 4.3 Continuous and Batch-Galvanizing Process

Attribute	Continuous Galvanizing	Batch Galvanizing
Zinc-coating thickness	10–35 μm	35–250 μm
Spangle	Light and even	Variable to large
Expected life span	10–15 years	30–150 years
Economy	Very economical	Economical
Limitations	Coils and wire	All shapes
Warping from heat	None. Sheet is levelled	Need to design around. Can warp thin plates and assemblies.
Defects – strings and patches	None. Continuous process should avoid	Possible. As the pieces are removed from the tank strings and splotches can appear.

FIGURE 4.41 Hot-dipped galvanized defect.

In many industrial applications, this would be fine, but for an exposed architectural bracket, it is not always acceptable. To achieve success with batch galvanizing it is important to establish tighter levels of acceptability if the end result is an architectural feature. Most batch-galvanizing companies understand this and, if aware, will assist in achieving success. If not, then another firm should be used.

There are other zinc alloy coatings performed by the continuous galvanizing process. There is also, electroplating, thermal spraying, and sherardizing, which are discussed in Chapter 3.

Many of the alternate coatings were developed to arrive at better coatings for the application of paint since the majority of thin sheet is destined for the automotive industry and the building products industries where typical applications apply paint to the surface.

Typical galvanized steel will not receive paint and must be treated before paint adheres to the zinc layer. One process that improved paint adhesion is called galvannealed. The galvannealed steel undergoes an additional treatment where the metal is annealed in an oven to a temperature of 565 °C (1050°F) after galvanizing to change the outer layer of zinc to one of iron–zinc. At this temperature the iron in the steel diffuses into the zinc layer making it an iron–zinc compound. The surface is darkened and smooth. It will accept paint better than a galvanized steel sheet.

Other coatings that uses zinc in a similar manner to the galvanized process are galfan, galvalume, zinc–magnesium–aluminum and several more. These are all produced in the continuous-coating process rather than the batch-galvanizing process. Some offer or claim to offer better corrosion resistance than galvanized sheet. The hot-dipped batch process develops the most resistant surface because of the ability to achieve a thick outer coating of zinc, much thicker than the continuous methods.

DARKENED GALVANIZED STEEL

There have been several artistic applications where the galvanized steel has been darkened. Chapter 3 discusses several. The darkening process attacks the outer zinc coating, so the need for a thickened zinc layer is paramount.

There are several proprietary treatments for darkening the outer layer of zinc fasteners. Both cold and hot processes can be used. The difficulty is the activation of the surface considering the galvanized coating is designed to resist corrosion. Nearly all of these processes involve acids, which initially attack the zinc surface. A ferric chloride solution, for example, is one that will darken zinc quickly. In solution, this highly corrosive chemical becomes muriatic acid, a dilute form of hydrochloric acid. Other solutions use boric acid and sodium acetate along with other compounds. In all cases, if you attempt the darkening of galvanized, do it in a well-ventilated space, wear protective gear, and do not breath the fumes. Collect the runoff and dispose of it according to regulations covering disposal of metals and acids.

On the hot-dipped batch-galvanized surface, the zinc coating is not consistent. Some areas are thicker than others. On flat plates, there is a difference in zinc composition around the perimeter edge. Cooling rates are different in the batch-galvanizing process so some areas may have more iron than others. These may go dark initially but the surface can be dusty. Once rinsed the surface is left with a mottled appearance.

As the surface further weathers, zinc hydroxide, white in color, appears against the contrasting dark background. This particularly happens around the edges. Figure 4.42 shows the effect of weathering after a short period of time.

The continuous hot-dipped galvanized coating has an even layer of zinc as the cooling rate is more defined. The difference is, the coating of zinc is thinner. The black or darkened surface is more

FIGURE 4.42 Darkened galvanized steel used on the Diesel Building in Chicago.

consistent. Still, zinc hydroxide will form along the edges where moisture from condensation will collect.

If it is simply a light darkening or aged look, vinegar is a simple approach. Even on large surfaces vinegar can be applied to a clean, dry galvanized surface and rinsed. This dulls the surface down. Vinegar, or more specifically, the weak acetic acid in vinegar, will slightly etch the outer zinc surface. It can also enhance the spangle, making the edges of the crystal more defined.

A technique that uses vinegar to age galvanized steel involves mildly abrading the surface while applying vinegar. White ScotchBrite™ should be used. Follow with denatured alcohol wipe.

Another method used to darken galvanized steel requires immersion of the galvanized steel in a heated solution of the following:

	Heat the solution to 60°C (140°F)
1 liter	clean tap water
5 cc	phosphoric acid
1 g	zinc nitrate
0.25 g	sodium fluoride
0.50 g	nickel chloride

This will quickly darken a clean, degreased galvanized surface.

There are Parkerizing treatments that create a zinc phosphate on the surface. This surface is even and darker than aged galvanized. Additionally, there is improvement in the corrosion resistance.

CHAPTER 5

Available Forms

A round man cannot be expected to fit in a square hole right away. He must have time to modify his shape.

Source: Mark Twain, More Tramps Abroad, Volume 2, 1897

INTRODUCTION

The various forms of zinc available to art and architecture are not as broad as those available for stainless steel, copper, and aluminum. The reason for this lies with the attributes of zinc and how the marketplace has come to perceive zinc. If galvanize steel, however, is included in the available forms you expand the breadth of zinc to include the forms and dimensions of steel products.

There are two main driving forces at play. Economy and strength. The economy of steel wrapped in the protective layer of zinc Increases the lifespan of the less expensive steel. Pure zinc lacks strength. Steel adds the missing strength.

Zinc is an excellent cladding material as long as what it clads delivers the mechanical properties zinc lacks. For this reason, we do not see zinc piping or tubing, structural angles or I sections. However, we have many examples of zinc clad by means of galvanizing, steel piping, tubing, angles, channels, and all the other structural steel components.

Table 5.1 compares a few of the attributes of zinc to other metals used in art and architecture.

You would think the lower melting point of zinc would make it ideal for cast elements. The melting point of zinc is significantly lower than other metals. The metal flows well and the softness makes cleanup of the cast surface easier. Figure 5.1 is test section of a cast fence design. The molten zinc was poured onto sand with stones placed to create a fence.

TABLE 5.1 Comparative Attributes

Attribute	Zinc	Copper Alloys	Aluminum	Stainless Steel	Weathering Steel
Tensile strength	Low	Moderate	Moderate	High	Very high
Bearing and wear	Good	Excellent	Excellent	Good	Good
Hardness	Low	Low	Low	High	High
Forming	Good	Excellent	Excellent	Good	Good
Stamping	Excellent	Excellent	Excellent	Good	Poor
Machining	Excellent	Excellent	Excellent	Good	Poor
Melting point	Low	Moderate	Low	High	High
Ease of casting	Excellent	Good	Good	Difficult	Difficult
Mechanical finishing ability	Good	Excellent	Excellent	Excellent	Poor
Patination	Good	Excellent	Poor	Poor	Good
Weldability	Good	Good	Good	Excellent	Excellent
Oxide runoff	None	Appreciable	None	None	Appreciable
Electrical conductivity	Good	Excellent	Excellent	Fair	Poor
Corrosion resistance	Excellent	Excellent	Excellent	Excellent	Excellent
Vibration dampening	Good	Excellent	Excellent	Good	Good
Magnetic	No	No (some alloys)	No	No (some alloys)	Yes
Recyclability	Excellent	Excellent	Excellent	Excellent	Excellent

The low melting point is another reason why the galvanizing of steel is so prevalent. The molten zinc is maintained for days as steel parts and assemblies are immersed into the bath to be coated. Figure 5.2 compares the melting point of various metals.

The melting point of tin and lead are lower than zinc. This makes the tin-lead solders ideal for joining zinc articles. Zinc is sacrificial to most other metals with the exception of magnesium, but the ratio of areas plays a part usually when zinc seaming undergoes soldering. Chapter 7 discusses this in depth. Essentially the area of zinc on most soldered assemblies is significantly larger than the solder joining material. The ratio of areas in most cases plays a major role in determining if a metal is going to corrode in a bimetallic coupling. What is critical is to clean the joint of all fluxing material once the solder joint is complete.

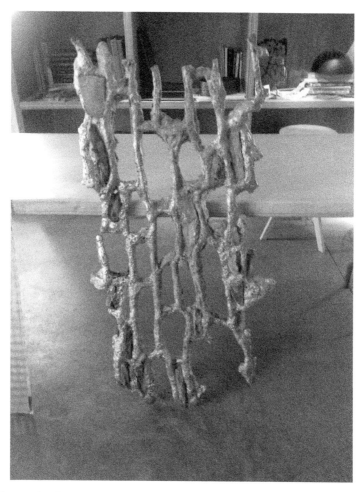

FIGURE 5.1 Rough cast zinc fence prototype.

Today zinc alloy is a common casting metal for small yet strong and dimensionally accurate parts. Many of our household appliances, keys in our pocket and parts in our automobile are cast from the zinc – aluminum alloys. In Chapter 2, Figure 2.3, shows the approximate overall percentage of zinc usage in various industries. Zinc castings, use more zinc by an order of two, than what is used in art and architecture today.

From a historical perspective, zinc once rivaled bronze as a cast medium for statues. In the late 1800s and early 1900s zinc statues, statuettes, plaques, and monuments were cast from zinc. Many were plated with copper to resemble bronzes. Zinc statues were sold as "white bronze." They were more economical than copper alloys and could be cast several times in the same mold. Refer to Figure 5.3.

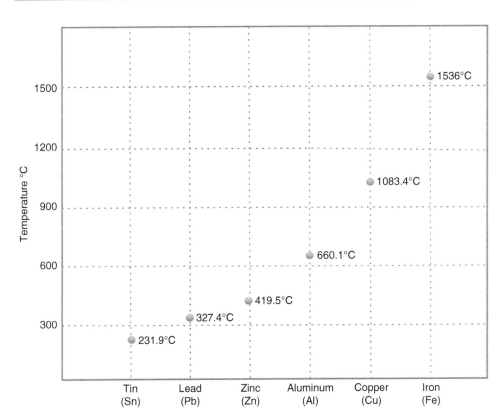

FIGURE 5.2 Melting point in °C of various metals.

WROUGHT FORMS OF ZINC

In art and architecture today, the most widely used form is the wrought zinc form. The wrought form of the metal, sheet, plate, wire forms have been around for centuries in one form or another.

Wrought Forms
Plate
Sheet
Foil
Wire
Extrusions

Mansard roofs of thin zinc defines the skyline of Paris, and wrought forms of zinc were in widespread use in Berlin through the 1800s as were castings and stamped plates. However, the

Wrought Forms of Zinc 149

FIGURE 5.3 Cast zinc statuette.

utility and acceptance of zinc as an architectural cladding material, particularly in North America, was limited until adjustments to alloying came into widespread architectural use, in the 1960s.

Prior to this time, the rolling of zinc was more cumbersome that rolling of other metals. Zinc had to be rolled hot because it was brittle. A process called *pack rolling* was used to create sheets of the metal. Pack rolling was a slow operation. In pack rolling, the zinc ingots were melted and then cast into slabs using water cooled molds. The slabs would be heated to a range of 175°C to 250°C (347°F to 482°F) and rolled through a set of pressure rolls to reduce the thickness. Each pass would thin the metal out and elongate it into strips of zinc. These strips of zinc had significant mechanical directionality and would crack if formed parallel to the grain. These strips were cut down to a specific length and stacked into small groups called *packs*. These packs of zinc sheet would be further rolled at right angles to the original reducing roll direction that created the long strips. This would reduce the anisotropy in the metal and make it more workable as a cladding material. The alloy would not work harden or change in temper much, even though it went through significant cold working operations.

In the 1960s, this all changed. The alloying of zinc with copper and titanium became widespread. The alloying had been developed earlier in the late 1940s and there were alloys of zinc and trace amounts of copper that were in use, but it was this ternary alloy that was the real breakthrough. This alloying invention, coupled with the modern methods of continuously casting, made the sheet form of zinc a formidable architectural surfacing material. Building on its established market in Europe and later reintroduced into North America, zinc became more understood and recognized as surfacing material. The zinc–copper–titanium alloy is the most common zinc alloy used in art and architectural wrought fabrications.

It is interesting to note, that during this time period, the 1970s and 1980s, other gray tone metals such as lead coated copper and terne-coated stainless steel fell out of favor because of the perceived hazardous nature of the lead coating on the surface. Both of these sheet metals were in extensive use as architectural roofing materials throughout the 1950s, 1960s, and 1970s. The sheet metal industry in North America was well acquainted with working with these metals and supply houses stocked these metals for sale. Zinc was not known or well understood outside of galvanizing.

Other surfacing materials such as galvalume, which is a zinc–aluminum coating on steel, aluminized steel, and other coated steels never made real inroads in architecture.

The massive mining company, Cominco Ltd., began an early marketing campaign in North America to take advantage of the interest in a gray metal for architecture. Cominco is one of the largest mining operations in the world and one of the largest zinc mining companies. Soon after their marketing campaign, large European firms that specialized in architectural zinc came to North America. Zinc was positioned to make inroads in the sphere of architectural metals all across North America. Major marketing campaigns increased the awareness of zinc and major designers began to consider zinc for their projects. One of the first projects in North America, showing a major use of zinc as an architectural cladding was the Legislative Building of the Northwest Territories (see Figure 5.4). The building had zinc for the roofing system imported from Germany but presumably manufactured from zinc mined by Cominco Ltd. in the Northwest Territories. Large cast zinc panels are used to define and decorate the exterior walls. The louvers, interior column covers, and exterior walls are all clad in zinc. This incredible example of architecture demonstrated the beauty of the various forms of the metal zinc and the fortitude zinc can exhibit in tough climate conditions of the Northwest Territories of Canada. The zinc is all preweathered.

Domestic suppliers of wrought-zinc products were reinvigorated by the competition and the newly developing marketplace. The supply, education, and knowledge of working with zinc was reborn as many successful architectural projects gave proof of the viability of zinc.

The modern method of continuous casting is vastly more efficient than the old pack rolling methods. The zinc is cast into a cavity that is continuously moving while being cooled with water. Refer to Figure 5.5. The thickness of the hot plate rolled out this way is from 10 to 75 mm. The grains

FIGURE 5.4 Zinc clad Legislative Building of the Northwest Territories, Yellowknife. Designed by Taylor Architectural Group and Gino Pin.

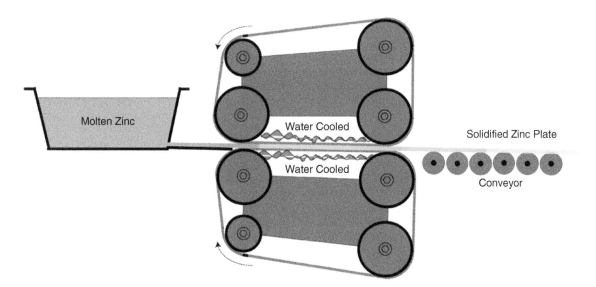

FIGURE 5.5 Continuous casting of zinc.

are finer and have less directionality. These plates are cooled further as they continually roll out of the machine and they enter a rolling mill that reduces the thickness to around 6 mm.

The plate further cools and goes through another set of reducing rolls, decreasing the thickness much further. The ribbon of metal becomes longer as the thickness reduces.

Zinc is available from many sources around the world; however, specific wrought forms used in art and architecture are available from a limited number of manufacturers. Limitations are imposed by the equipment used to manufacture the wrought form, such as the continuous cast equipment, the reducing rolls used to bring the thickness down and the mechanical limits of the metal itself. It is important to work with a supplier of the wrought zinc form to understand the limitations of production.

PLATE

Zinc is available in plate form. Zinc plate is considered in thicknesses greater than 2 mm (0.78 inches). The width is limited, as is the available finish from the mill source. Custom finishes are available from secondary sources. Above 2 mm thickness, the finish is coarse and rough. This is the nature of the hot-rolled plate. It is provided in an as fabricated natural zinc surface without preweathering or darkening.

Very rough plate arrives out of the continuous cast machine at approximately 15 mm thick as it is sent to the reducing rolls. Still as the rough plate is smoothed and stretched the surface inclusions and surface texture is coarse and unrefined. Figure 5.6 shows a typical zinc plate surface. Oxides, streaks, and small pits or inclusions can be seen in the hot-rolled surface. The finish is rougher than the smooth surface of sheet.

Plates are provided in a natural as cast finish. Figure 5.7 shows zinc planters made from 6 mm zinc plate. These have been cleaned by a secondary finishing process to even the surface out. The finish is typical of the more-unrefined plate surface.

FIGURE 5.6   Zinc plate surface.

Sheet and Coil 153

FIGURE 5.7   Zinc plate planter boxes.

Finished zinc plates are available in:

3 mm thickness
4 mm thickness
6 mm thickness

Rough zinc plates, lacking flatness can be obtained as thick as 15 mm.
The width of zinc plates is limited to 1105 mm (43.5 in.).

SHEET AND COIL

Rolled zinc sheet is the more common form of wrought zinc used in art and architecture. The zinc alloys containing small amounts of titanium and copper make up the sheet forms readily available to the market. The available thicknesses of this alloy have been defined by industry usage and the alignment with the manufacturers production processes.

Unlike the steel and stainless steel market where there is a vast array of thicknesses produced by the mill source, rolled zinc is similar to the copper rolled product market with a limited number of thicknesses being produced. Specific thicknesses could be achieved but there is not the demand. If the quantity was large enough, the supplier would most likely adapt to the request. Figure 5.8 are panels manufactured from 1.5 mm thick zinc sheet. The finish is Hunter™ over preweathered.

FIGURE 5.8   Hunter™ zinc on 1.5 mm thickness zinc.

It is advisable to verify the availability of one thickness versus another from the zinc manufacture. The following are the thicknesses normally available. These are nominal thicknesses and the manufacturing specifications allows for a tolerance of 10% of the thickness.

Zinc Sheet and Coil Thicknesses

0.7 mm	0.027 in
0.8 mm	0.032 in
1.0 mm	0.039 in
1.2 mm	0.047 in
1.5 mm	0.059 in
2.0 mm	0.078 in

These forms are available in nominal widths of 1,000 mm (39.4 inches). There are manufacturers that can push this to 1,105 mm (43.3 inches) in width. Figure 5.9 shows 1 meter wide zinc panels made from 2 mm thickness.

Coils are available in the various thicknesses and widths. Coils are common in roofing and roll forming operations where the coil is set in line to uncoil, level, and feed into a roll-forming machine to make the desired profile. The coils are provided in weights depending on how the receiving firm can handle them. Figure 5.10 shows approximate size and shape of zinc coils produced by the mill.

The rolled-zinc sheet surface is prepared in a number of ways. The zinc can be provided in the bright, natural finish but the typical supply from the mill source is in one of a variety of treated surface finishes. The matte-gray color tone we associate with zinc used as facades and roofs is produced in a preweathering process. All the major manufacturers of the rolled-zinc sheet and coil produce a preweathered surface.

FIGURE 5.9 One meter wide preweathered zinc panels. 2 mm thickness.

Over the years variations of this basic weathering finish have developed as the manufactures seek to diverge in their product offering. Some of the preweathered finishes are only available in sheet sizes while others are available in coil. Figure 5.11 show variations of the preweathered finish from different manufacturers with a couple of natural finish, bright samples, in the middle to demonstrate the contrast.

These preweathered finishes are not applied like paint or plating but are chemical reactions with the zinc surface. The finishes are somewhat different from one manufacturer to the next. Each manufacture of wrought zinc influences the way the surface looks on close inspection. The grain, the color and the reflective sheen are different for different manufactures of sheet zinc. This is due to the casting process equipment, the cooling water, the quality and finish on the reducing rolls and

Chapter 5   Available Forms

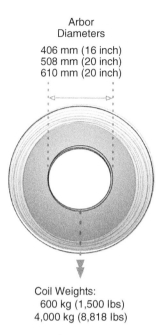

FIGURE 5.10   Examples of coil weights and arbor diameters.

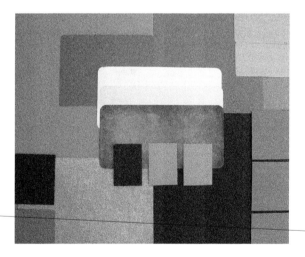

FIGURE 5.11   Variations of the preweathered option from various manufacturers.

the preweathering treatment undertaken. Some treatments use a phosphoric acid to create a zinc phosphate on the surface while others create a thin zinc carbonate.

Additionally, different batches from the same manufacture can be slightly different in color and tone. The preweathering process is a chemical reaction on the zinc surface and is subject to slight variations as the chemical reactions occur. Variations on the surface, temperature, chemistry strength, and time are just a few of the variables that play a role in arriving at the initial preweathered color.

Many manufacturers apply a light, organic clear coating on the surface of the metal. This is to resist staining and fingerprinting during storage and handling. It is not visible to the eye and

FIGURE 5.12 Dome with Roano™ patinated shingles.

will come off if wiped with a mild solvent. It is intended to wear off over time and exposure to the atmosphere.

There are some tinted finishes available from specific manufacturers of rolled zinc using mineral dyes in a thin clear coating to add soft colors to the surface. These semitransparent films are applied over the preweathered surfaces. These coatings are very thin. They hold up well and zinc is an excellent base for painting with certain paints, but their thinness would suggest they will eventually deteriorate in ultraviolet light and harsh exposures.

There are several patina finishes available from secondary suppliers. The patina finishes are only provided on sheet forms and to a lesser degree, the plate forms. These are chemical reactions on the surface, reference Chapter 3, and create a thickened layer of corrosion products that add protection as well as color to the underlying zinc. The patina finishes are available on sheet. Figure 5.12 is an image of a dome with the pre-patinated, Roano™ patina on shingles.

ZINC FOIL

Zinc is available in foil forms. Zinc foil is made by electroplating on an aluminum drum, then stripping the thin zinc off after it cools. It is further passed through special reducing rolls that squeeze and stretch the soft metal to the desired thickness. The high-purity form does not undergo cold working to any great extent and is soft and malleable. Thin zinc foils cut easily with hand shears, even household scissors.

Zinc foils are used in shingle roofing and other applications where corrosion resistance and ease of forming by hand is desired. Zinc foil is used as laminated tapes as well. The available thickness on the pure zinc foil is 0.2 mm but zinc foil as thin as 0.020 mm is advertised.

The zinc alloys are not commonly produced into foils, rather commercial pure alloy foils are normally available. The purity ranges available are 99.9%, 99.98%, 99.99%, and 99.999%. These are very soft and can be easily hand folded to fit contours.

EXTRUSION

Zinc can be extruded. The cross section is kept small, less than 5 mm, because it is so weak as it extrudes hot. Large sections collapse. Tube forms collapse. For a period of time it was extruded as U-shaped came for stain glass edging work. Subsequently it is rolled formed into small sections rather than extruded. Sources for extruded zinc are not readily available. Figure 5.13 shows an image of the zinc panels on the interior of the Catwalk Lounge of the Peninsula Hotel in Hong Kong. These panels have custom formed zinc came used at the joints between the 1 mm zinc.

FIGURE 5.13 Peninsula Hotel Honk Kong. Zinc panel with zinc came at the joints. Designed by Philippe Stark.

TUBE AND PIPE

Zinc, drawn or extruded into tubes are not readily available in the marketplace. Most, from an art and architecture context are made from sheet material. The availability, strength, and low cost of galvanizing steel places pure zinc pipe and tubing at an industrial disadvantage. It is much easier and significantly more economical to produce the forms from galvanized steel.

Zinc drawn tubing is available in round or oval form. The cross sections are small, and the availability is limited. The diameters available are from 0.5 mm to 150 mm (0.02 in. to 6.0 in.) and wall thicknesses from 0.07 mm to 12 mm (0.003 in. to 0.5 in.). Lengths are short, typically no longer than 1800 mm (72 in.).

WIRE

Wire forms of zinc are readily available. Zinc wire is used extensively as source for protecting steel using thermal spray devices. Zinc wire is fed into a flame or plasma arch and atomized into fine particles. Reference Table 5.2.

Thermal spraying zinc not only adds galvanic protection to steel surfaces, but it also can be used to fill gaps and level a surface for painting. The surface produced is very porous and will accept paints.

Wire is produced from zinc rod that has been extruded from a cast alloy billet. The rod is further drawn through a series of dies in a continuous process that produces wires of various cross section. Different alloys can be cast and drawn in this manner.

The zinc alloy, commonly known as 85–15 for 85% zinc and 15% aluminum is a common alloy used for thermal spraying or metallizing of steel surfaces. Pure, 99.99% zinc is also used frequently. Other alloys, containing 2, 4, or 22% aluminum with zinc are produced in wire specifically for the metallizing industry.

ROD

Rod forms of zinc are often used in various aqueous applications where they are provided as sacrificial anodes to protect other metals in proximity. Water heaters, marine, boilers and other articles where aqueous solutions can corrode the piping, or the walls of the tank commonly use zinc rods to act as sacrificial anodes. Zinc rods and bars are placed in the ground and attached to steel sculpture to act as a sacrificial anode to the sculptural form.

The rods are extruded in various diameters. The lengths are limited. As the metal is extruded the rods are cut to lengths that are used in industry. Table 5.3 show the nominal size of rods available in zinc. Dimensions will vary depending on North American or European supply.

WIRE MESH

There are no producers of standard mesh made of zinc. There is no reason other than demand. To set up the machines to weave zinc wire into a screen or mesh would require a quantity and this, in turn, a market. There are a variety of galvanized steel screens and meshes readily available, but there are none stocked or offered made of zinc wire.

EXPANDED METAL

Another type of mesh, expanded metal, can be made from zinc sheet. Again, the market is not there and thus no expanded metal producer stocks expanded zinc. It would need to be a special order but

TABLE 5.2 Wire Diameters Available in Zinc Alloys

mm	in.
1.45	0.057
1.62	0.064
2.0	0.079
2.3	0.090
2.5	0.098
3.0	0.120
3.2	0.125
4.76	0.187

TABLE 5.3 Available Zinc Rod Diameters

mm	In.
6.35	0.25
9.25	0.37
12.70	0.50
15.87	0.62
19.05	0.75
20.95	0.87
25.40	1.00
31.75	1.25
38.00	1.50
50.80	2.00

the criteria for making expanded metal from zinc sheet would be the same as it is for other metals such as copper or aluminum.

Expanded metal is readily available in galvanized steel. Often, the steel is expanded and then sent to the galvanizing facility to coat the surface and pieced edges created by the expanding process. This way all the surfaces are coated with the protective zinc layer.

There is nothing to prohibit creating expanded zinc sheet. The edges would be homogeneous with the balance of the metal and fabrication of the expanded zinc would not pose any particular issue with the machinery used.

PERFORATED ZINC

Perforating zinc sheet is available in both custom perforations and standard grids. Zinc perforates similar to aluminum with very low energy and force requirement. Figure 5.14 shows several examples of perforated zinc.

Similar to aluminum, copper, and stainless steel, perforated zinc is homogeneous, and all edges and surfaces are zinc alloy. Whereas, in the case of galvanized, often the metal is perforated after galvanizing and the zinc is required to expend sacrificial protection to the cut edge of the exposed steel. Post galvanizing of perforated steels may result in filled or partially filled holes. When post galvanizing perforated steel, air blow the opening to remove the still molten zinc.

Thin galvanized steel can be perforated, and the zinc will offer protection to the edge up to a point. Generally, in most mild exposures, galvanized steel sheet less than 3 mm will be protected. Keep in mind, the continuous galvanizing process produces a sound coating on the surface of steel, but it is thin. There is less zinc coating the steel.

Perforated zinc can be patinated either before or after the perforation process. The zinc offers a sound, corrosion resistant base material. There is no work hardening when perforating so the sheet does not warp or curve as it can with other metals.

Figure 5.15 shows an example of patinated and perforated zinc.

FIGURE 5.14 Custom perforated zinc examples courtesy of ImageWall.

FIGURE 5.15   Black zinc bridge panel.

TEXTURED ZINC SHEET

Rolling texture into zinc sheet adds stiffness and character into the flat surface. Zinc readily accepts embossed patterns achieved by rolling sheets through pattern rolls. Figure 5.16 shows a hammer-tone pattern induced into the surface by rolling the thin sheet through a set of matching pattern rolls.

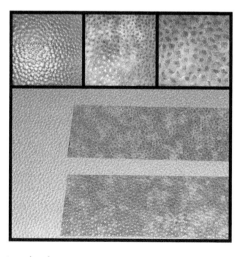

FIGURE 5.16   Hammer tone texture in zinc.

All embossed patterns have two sides. One with the pattern going in and one with the pattern going out. In Figure 5.16, the upper-left image is the opposite side achieved with the lower image. Patinated zinc can be textured in a similar manner. The embossing does not damage the patina.

The advantages with the embossed surface is it stiffens the sheet and it can hide small imperfections and damages that may happen during use. Light reflecting from the surface is slightly scattered from all the surface indentations. Oil canning is not as visible. Figure 5.17 shows custom embossed zinc surface at a grazing angle. The surface appears very flat as the texture stiffens the thin metal.

FIGURE 5.17 Custom texture on zinc. Courtesy of Rimex.

ZINC ORNAMENTATION

In the latter part of the 1800s, architecture sought ways of adding ornamentation to our buildings. In Europe and later in the United States, this practice included adding decorative ridge features, gargoyles, griffins, and other decorative features. Spun balls and spun baluster forms were used to decorate the edges and surfaces of Victorian-style architecture. Figure 5.18 shows an example of this. The decorative ridge treatment were stamped of zinc and spun finials were made of zinc. Usually these would be left to weather naturally and left unpainted.

Zinc is well suited for stamping, and many early zinc sculptures were made by stamping warm zinc sheets into molds. There are companies that still stamp zinc sheet into wooden or hard urethane

FIGURE 5.18　Vaile Residence. Built in 1880 and restored in the 1960s. Independence, Missouri.

FIGURE 5.19 Stamped zinc forms.

molds. You can purchase sections of stampings and assemble them into ornamentation. Figure 5.19 shows a stamped zinc leader in the form of a fish head. The stamped pineapple form finial is assembled from three sections stamped from the same die.

CAST

Second only to galvanized coatings of steel, cast zinc articles are used in everyday household products, transportation, tools and toys. Other than the rolled zinc roofs in Paris and Belgium, casting was the form zinc was most frequently found in art and architecture. In the middle to late 1800s, zinc cast statues and statuettes were common.

FIGURE 5.20 Dough boy made by slush casting of zinc, then copper plated.

The ability and ease of plating zinc was developed the mid-1800s. Copper plating of zinc became common. It was more economical to make the statue and statuettes from zinc and copper plate. In North America, many early-1900 statues were plated in copper. Zinc was less expensive to cast, and many replicas could be produced from a single mold. Figure 5.20 is an example of the American Dough Boy sculpture that proliferated around the United States after World War I.

SLUSH CAST

Many of these early statues were cast using a method called slush casting. Zinc was ideal for slush casting. With low melting point and good fluidity, zinc would be "slushed" into a mold made of refractory material then quickly poured back out. This would be repeated until a given thickness was arrived at. One drawback with this casting method was the weight of zinc. Pouring it in and back out of the mold was difficult on large castings.

Slush casting began in the early 1800s with lead as the metal being cast into iron molds to make toy soldiers. The molten metal was poured into a two-part mold and quickly poured out, leaving a hollow lead toy soldier. Zinc, another low-melting-temperature metal, was used to make toys but also larger statuettes and household items. The use of slush casting peaked sometime around the 1930s. There was a proliferation of slush cast "dough boys" after World War I. Figure 5.20 is one such monument. This particular one is in front of the Leavenworth, Kansas courthouse. Many others like it can be found in small towns and cemeteries across the United States.

For slush casting, the dies and molds are all custom made. The zinc is ordered in the alloy desired and arrives as small ingots to be melted down. The molten zinc is poured into the mold and then quickly poured out leaving a thin layer of solidified metal on the sides of the mold.

Slush casting zinc for architectural use today is not practiced.

DIE CAST

Die casting of zinc is in common use across many industries. Zinc alloys can produce dimensionally accurate shapes with little post cleanup. Zinc die castings are small in overall size and are set up for production runs of several hundred or thousand parts. The strength and durability of the metal is superior to plastics, so zinc die castings find their way into many home appliances with moving parts such as blender motor housings, gears, and carbonators for small engines.

Die casting came into widespread use when gravity cast was replaced by pressure casting where the metal is rapidly injected into a steel mold. Zinc is especially suited to this form of casting when production runs of small, accurate parts is demanded. Zinc pressure casting uses a "hot chamber" where a hydraulic piston is immersed in the molten metal and is used to drive the molten zinc into the die. The die is made of two halves and is steel. It receives the molten metal and once the metal has solidified, the two halves of the die open and the finish part is removed. The steel die then closes, and the process is repeated.

For die casting, the most common alloys used are the Zamak series containing aluminum, magnesium, and copper. The alloy Z33520, Zamak 3 is the most commonly used alloy by the die and spin casting industry in North America. Z35531 alloy, also known as Zamak 5 is widely used throughout Europe die-cast operations. Table 5.4 shows the UNS alloys and the common names of these zinc–aluminum cast alloys.

The Zamak family of alloys was developed by the New Jersey Zinc Company in 1929. The industry wanted a strong, durable inexpensive die cast metal for the burgeoning automotive industry. Prior to this time, there were no real standards. Inexpensive zinc castings were made for the toy industry and other home products. It was called "pot metal" or "monkey metal." It acquired this early name from the collection of scrap into a bucket or pot in the small metal shop. It might get melted down and cast into some temporary tool or bracket. There were no controls.

These stronger alloys of the Zamak series, designed for specific characteristics, were created as aluminum came on the metallurgical scene in the early part of the century and began to compete with the zinc as a casting metal. Aluminum has a higher melting point but still lower than steel or brass and aluminum was stronger than the original zinc castings, which were prone to cracking and which lacked sophisticated metallurgical consistency.

Once this family of zinc–aluminum alloys were created the zinc casting industry was able to take back and expand the die-casting market, with stronger, more durable yet affordable castings. These alloys cast at lower temperatures than aluminum and had good flowing characteristics into molds. Molds could be made from materials that would allow repeated castings.

TABLE 5.4   Common Die-Cast Alloys of Zinc

Alloy	Common Alloy #	Other reference names	Aluminum	Magnesium	Copper	Distinguishing Attribute
Z35541	#2 Zamak 2	ASTM AC43	3.5–4.3	0.2–0.5	2.5–3.0	Highest tensile strength, good creep resistance
Z33520	#3 Zamak 3	Mazak 3 ASTM AG40A	3.5–4.3	0.2–0.5	0.25	Good strength, creep resistance, and fluidity
Z35531	#5 Zamak 5	ASTM AC41A	3.5–4.3	0.3–0.8	0.75–1.25	Hard alloy. Strong and good creep resistance. Most widely used die-cast alloy in Europe
Z33523	#7 Zamak 7	ASTM AG40B	3.5–4.3	0.03–0.05	0.25	High ductility

SAND CAST

Zinc is cast in the traditional sand molds used for casting bronzes. Zinc is easier and less costly to cast than other metals because of the low melting point and the softness of the metal allows for less rigorous finishing. The drawback to large zinc sand castings lies in the weakness of the metal and the tendency to undergo creep. More compact castings are made, then assembled into larger pieces.

Zinc is not the common cast metal for art or architecture. It is more a condition of availability and specific taste. Zinc sand casting is not as available to the market because of a lack of understanding. Most foundries work in copper alloys or steel. They understand these metals and understand how to create the finishes and patinas.

Zinc casting was widely used in Europe in the late 1800s and early 1900s, but the metal as a cast form for art and architectural embellishment has all but disappeared. Figure 5.21 shows a large cast zinc sculpture used in a fountain. The color is a deep blue gray tone. This was cast nearly 100 years ago.

The constraints with casting zinc reside in the strength of the metal when cast into large sections. The tendency for the material to creep is a crippling mechanical behavior that limits heavy sections set as cantilever extensions. The zinc compact hexagonal crystal allows slippage to occur at room temperature. This self-annealing behavior is overcome in the alloying used for rolled zinc products but for large castings it still must be considered in the design. Figure 5.22 shows a crack that has formed in the outstretched leg of one of the horses.

Foundries are concerned with the dust and the fumes from zinc and the health hazard this can present. Chapter 1 briefly discusses the zinc fever that can occur when breathing zinc fumes.

FIGURE 5.21 Cast zinc sculpture.

FIGURE 5.22 Crack from creep stress in outstretched leg.

Patination and finishing are not practiced on zinc at foundries as it is with copper alloys. The resulting surface of a zinc casting does not require the coatings copper alloys require to protect them from interacting with the environment. However, it is good practice to wax a zinc sculpture to reduce staining and make it easier to clean the sculpture, but it is not required.

Today you see more aluminum and even stainless steel sculpture castings than zinc sculpture castings. These metals require more energy and finishing effort, but they possess the strength to withstand the forces imparted by wind, gravity and atmospheric corrosion. Table 5.5 shows some comparative attributes of various metals considered for artistic casting.

Zinc casting would be a good choice if the object to be cast is more compact. Large sculpture with extended or cantilevered sections would not be suitable for zinc over the long-term. Cracks will

TABLE 5.5 Comparison of Different Metals Considered for Casting

Parameter	Zinc	Copper Alloy	Aluminum	Stainless Steel
Energy required	Low	Medium	Low	High
Fluidity	Excellent	Good	Good	Fair
Rapid die casting	Excellent	Good	Good	Poor
Mold cost	Low	Medium	Medium	High
Metal cost	Low	High	Medium	High
Weight	High	High	Low	High
Creep/Mechanical strength	Poor	Good	Good	Good
Detail ability	Excellent	Excellent	Fair	Excellent
Corrosion resistance	Excellent	Good	Excellent	Excellent
Protective coating	None	Required	None	None
Patination ability	Limited	Vast	None	None

develop over time as the metal is stressed. Reference Figure 5.23. This sculpture has been subjected to wind, freeze–thaw forces, and gravity. But it is more compact and thus does not see conditions that can lead to creep. Once a crack does begin to form from creep, it is difficult to repair.

ZINC POWDER

Zinc powders are used in many applications unrelated to art and architecture. Zinc powder is created from condensing zinc vapors by shock chilling in nitrogen. Tiny spherical particles, almost like dust is made this way and collected for inclusion in everything from paints to cosmetics.

In paints and varnish the fine powders are used to equip them with a level of sacrificial coating. These metal powders also increase the density of these paint films making them viscous and adding to the barrier aspect of the coating.

Zinc rich coatings are used frequently for protecting steel. To be considered zinc rich, the organic coating is approximately 90% zinc when the film has dried. The organic paints that carry the zinc powder are often epoxy, but polyester and chlorinated rubber coatings are sometimes used. There are inorganic, silicate-based coatings that also incorporate zinc powder.

Zinc powder is used in flame spray applications; however, the preferred method today for thermal spray coatings is wire feed. Zinc powders are used in flame spray apparatus equipped with a hopper-type feed. The hopper has to be filled frequently, whereas a wire feed apparatus is more effective and is usually continuous fed.

FIGURE 5.23   Smaller compact zinc sculpture of a Shepard boy.

Zinc powders should be stored in a dry place. They can become explosive and ignite if they are allowed to get damp. The rapid formation of zinc hydroxide on pure zinc powder can cause an explosive reaction to occur.

CHAPTER 6

Fabrication

I am always thinking about creating. My future starts when I wake up in the morning and see the light.

Source: From Davis, Miles, and Quincy Troupe. Miles, the autobiography. New York, © 1990, Simon & Schuster.

WORKING WITH ZINC

In the hands of the fabricator, zinc is similar to most of the other metals used in art and architecture. There are a few nuances that distinguish zinc or define it in the eyes of a fabricator. In general, though, zinc is near the weight of copper or steel of similar dimension. It has a softer feel, it bends easier when flexed in the hand, and the edge is not as sharp and defined as stainless steel. Zinc's appearance in the natural, clean, unweathered state, is indistinguishable in most lights from aluminum. Once the oxide begins to form, it darkens and takes on a gray tone, giving it a tin- or lead-like appearance.

Major Attributes of Zinc from a Fabricator's Perspective
- Ease to form
- Ease to stamp
- Prototypes are simpler to manufacture
- Can melt down and remake
- Does not spark
- Self-lubricating

- Machining is fast
- Ease of casting
- Good fluidity for detail casting
- Accurate casting
- Low temperature
- Good impact strength
- Simple to polish
- Good abrasion resistance
- Electromagnetic interference shielding capability
- Nonmagnetic

Whether cast or wrought forms of zinc, fabricators must make a few adjustments to account for the subtleties of zinc. The main nuances a fabricator must consider when working with zinc are:

- Protecting the surface
- Anisotropy
- Temperature

STORAGE AND HANDLING

Zinc is susceptible to stains from moisture collecting on the surface. The staining is a white, crusty development of mostly zinc hydroxide. Figure 6.1 shows what this stain looks like when it develops from moisture entering between zinc sheets. If moisture is allowed to reach between stored sheets of zinc, a stain can develop. On the natural zinc surface, one that is not preweathered, the stain will happen quickly. It can happen on galvanized surfaces as well. This is commonly called *white storage stain* and it is a white zinc hydroxide, a thickened oxide layer that forms quickly when moisture is present and air is restricted.

The rolled-zinc sheet should be protected with plastic film, both the preweathered zinc and the natural zinc. This will offer some protection, but care should be taken to prevent moisture from reaching the edge and wicking onto the surface. The film will also offer some protection during the handling of zinc sheets during fabrication. Zinc is soft and the surface can scratch from other metals, sand, slag, and other hard substances that are dragged over the surface. Figure 6.2 shows a perforated zinc wall with the plastic protective layer. The layer is peeled after installation.

When you receive rolled zinc, even when coated with a thin plastic, it often has a mill oil applied. This oil is imperceptible both visually and to the touch. It helps to protect the surface from moisture in the early days of handling and exposure. On natural finish zinc, in particular a thin sheet, the oil can be heavier, similar to that on cold-rolled steel.

Zinc castings should be stored in dry spaces protected from condensation and foreign substances that can collect on the surfaces. Zinc castings should be protected from rubbing and shifting on other castings or where abrasive materials can come in contact with the surface. Because of the relative

FIGURE 6.1 White storage stain on the left edge of a patinated zinc surface.

softness of this metal, it can be easily gouged and scratched. Protective wraps or foam separators, preferably non–moisture absorbent, are recommended.

Zinc-coated steel using the hot-dipped galvanizing method is also prone to handling damage as the heavy parts, often awkward in shape, bump into or are set down onto other parts. It is very important to use wood or other separation to keep from dragging the heavy parts over one another. See Figure 6.3 for damage caused by mishandling exposed galvanized steel. These are exposed brackets, and when they were removed from the molten zinc bath, they were set on other parts, creating a mark. The corrosion resistance of the zinc is not affected.

This type of damage is not repairable. It might weather down a bit, making the damage less apparent, but it will be visible for the duration of the part's life.

176   Chapter 6   Fabrication

FIGURE 6.2   Zinc wall with protective plastic peel coating. Removed after installation was complete.

FIGURE 6.3   Damage to galvanized steel during storage and handling.

CUTTING ZINC

Zinc can be cut with all standard methods used to cut other metals. The force needed is less and cutting with abrasives is simple and efficient. Zinc is nonsparking, so cutting with high-speed abrasives does not generate sparks as can occur with the steels and stainless steels. When cutting with abrasives, zinc particles can be larger, as the harder material takes out sections of the softer zinc metal. The spray of these particles is not molten like steel. It can be hot from the friction developed but not to the degree of steels or stainless steels, where they might bond onto the surfaces.

SHEARING AND BLANKING

Zinc sheets normally used in art and architecture, thicknesses from 0.7 mm to 1.5 mm (0.027–0.060 in.), pose no difficulties with shearing or blanking when using conventional shear or blanking equipment. Rolled-zinc sheet of the convention used in art and architecture is tougher and more resilient than pure zinc sheet, but the shear strength for zinc is still low. Table 6.1 shows comparative metals used in art and architecture and the approximate shear strength.

SAW CUTTING

Saw cutting with toothed cutoff saws is a common method of addressing zinc plate, bars, and shapes forms. Use coarse-toothed blades or coarse-toothed bandsaw blades to cut zinc forms of thickness. Saw cutting should be used on zinc forms of sufficient thickness to avoid distortion at the cut. Thin zinc sheet or shapes of thin zinc can be cut on bandsaws or handheld saws when properly supported. Coarse-pitched blades are preferred, and speeds can be high. For circular saws, coarse-pitched

TABLE 6.1 Shear Strength of Various Metals

Metal	Shear Strength (psi)	Shear Strength MPa
Aluminum	8,000	55.2
Rolled zinc	14,000	96.5
Copper	28,000	193
Copper alloy (Muntz)	40,000	276
Steel	43,000	296.5
Stainless steel	50,000	345

LASER

Zinc can be cut with a laser. Lasers are excellent and efficient cutting tools for cutting all zinc alloys up to thicknesses of 2 mm. Above 2 mm, some melting occurs on low-power lasers. Intricate shapes and detail can be cut into zinc sheets. A laser works by concentrating a highly focused beam of light onto the metal surface. Electromagnetic energy of the laser light is transformed into thermal energy within the thickness of the metal. The light is absorbed by the metal, and this energizes the electrons. The electrons are accelerated by an energy field within the crystal lattice of the metal, and this generates heat. At the point where the light strikes, the heat is so high that the zinc melts or vaporizes. The amount of heat generated is based on the light absorption of the material being cut. Once it pierces the metal, the cutting action begins. The beam of light moves along the contour, melting the metal, while a jet of gas blows the melted metal downward, leaving a kerf cut not much wider than the beam itself. Nitrogen atmosphere works well for clean, fast cuts in zinc sheet or thin plate. Zinc absorbs light energy similar to steels, but with less energy needed to bring the zinc to the melting point.

The surface properties of the metal influence the optical behavior of the laser beam. The peaks and valleys of a diffuse surface can trap some of the light and speed up the process. Zinc sheet and plate can be cut with gas lasers (CO_2) and with fiber lasers. Fiber lasers are solid state, Nd:YAG (neodymium-doped, yttrium-aluminum-garnet) lasers that deliver the energy by fiber. This is a very efficient method of cutting zinc, steel, stainless steel, aluminum, and even copper.

Higher-powered lasers are needed to cut thicker metal. A 2000-watt laser can cut 4-mm-thick zinc adequately without leaving a burr or edge. The edge produced on a laser should be clean and free of heat discoloration and burrs from the cutting process. Heat tint is usually not an issue with laser cutting when using a nitrogen atmosphere.

Lasers are CNC-controlled cutting devices. They are very efficient cutting tools and can be programmed to cut intricate shapes in both two-dimensional surfaces and three-dimensional parts. If there are a lot of piercings in a given design, some shaping may occur in the surface as heat is absorbed by the metal. Additionally, there can be discoloration along the area where a laser drill or piercing occurs. Waxing the sheet or plastic coating the sheet can resist some of the discoloration that occurs around the cut as it keeps splatter from molten metal from adhering to the surface.

The steel slats used on the beds of most laser-cutting machines will not contaminate the zinc. Any steel particles that would get embedded into the zinc would be protected by the surrounding zinc due to the massive differential in ratio of areas. However, the steel slats can scratch the softer zinc if the sheets are not handled on and off the bed with care. The slats can also scratch any paint coatings applied to the reverse side of zinc sheet, rendering their protective ability suspect.

PLASMA

Plasma cutting involves the development of an electrical current from the conductive metal part being cut to the plasma nozzle. A jet of compressed air is passed through the small orifice on the tip of the nozzle, and as the electrical current is generated from the metal part, the jet of air is superheated to a plasma state. Temperatures in the high-velocity plasma jet can be as high as 22,200°C, as this highly charged gas melts the metal and blows it away.

There is not a lot of heat transferred to the zinc but if there are a number of piercings, you can expect some warping of the thin metal. The high-definition plasma-cutting systems reduce the heat-affected zone and produce a fine cut line in zinc sheet. This hot jet of compressed air melts the metal being cut and blows it away, leaving a cut.

The conductivity of zinc enables the cutting with the plasma torch; however, the fumes generated are hazardous and should not be inhaled. See Chapter 1 on the discussion of zinc fumes and *zinc fever*. This fuming hazard also occurs when cutting galvanized steel with a plasma torch. It will cut well, but the fumes generated are toxic and should be avoided.

The handheld plasma cutters and some of the less-sophisticated systems will have a larger kerf cut, rougher edge, and more oxidation on the edge. The *kerf* of the cut, which is the term given to describe the edge or cut, is rougher than water or laser. There can be a redeposit of molten metal along the kerf. This redeposit is rough oxide that will need to be removed. For art and architectural projects, plasma-cutting zinc has been displaced by the high-quality and speed provided by fiber laser systems. High-definition plasmas are fast and efficient and can be considered for many projects involving cutting of zinc – in particular, those where post-surface finishing will occur.

Another benefit of plasma cutting is that it can be used in the field to trim and cut zinc. Using a straight edge or a template, plasma cutting large holes or shapes in the field is a viable option when trimming thin zinc sheet to fit a particular application. It is very important to avoid the fumes created by this cutting operation when performed in the field by handheld plasma-cutting devices.

WATERJET

Zinc sheet, plate, and castings can be cut using a waterjet. The waterjet stream with garnet will slice through even thick zinc plates and castings. Cutting zinc plates with waterjet is similar to cutting thick aluminum plates. Waterjet cutting is not as fast as laser cutting, but thicker materials and shaped forms can be cut more effectively with a waterjet. The water does not affect the zinc surface. There can be frosting around the initial piercing of the zinc similar to other metals. As the waterjet begins, the pierce it must drill through the metal, and this can leave a frost mark around the initial hole. Coming from the edge, the powerful stream of water acts more like a saw as the garnet in the stream of water saws the softer zinc away. Refer to Figure 6.4.

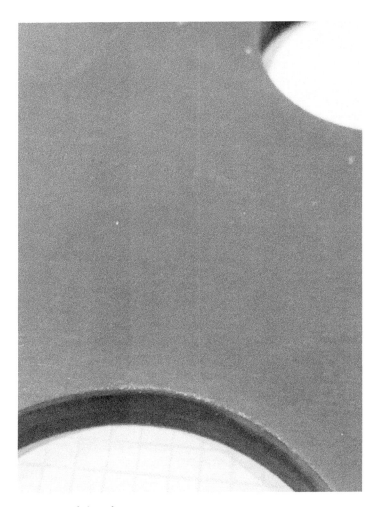

FIGURE 6.4 Waterjet piercing of zinc plate.

PUNCHING / PERFORATING / BUMPING

Zinc is readily punched to create indentations or perforations. The yield strength is low, and zinc puts up little resistance to modern CNC machines or roll-type perforating machines. Figure 6.5 shows a custom perforated zinc sheet with the Roano™ patina. The sheets were corrugated after the pattern was produced.

Zinc has a relatively low-yield strength, and cold shaping rolled-zinc sheet is easily accomplished as compared to other metals used in art and architecture because of two additional properties. Zinc ability to elongate under load and plastically deform to the die shape and zinc's

FIGURE 6.5 Custom-perforated ImageWall® and patinated zinc on Denver, Colorado, residential tower parking garage. Designed by the Beck Group.

self-lubricating character. Zinc has excellent tribological characteristics because it is soft and tends to smear at the interface with harder tooling.

Figure 6.6 shows a custom formed, punched, and bumped blackened zinc sheet. This custom cladding combined with the shaping simplicity of zinc to create this unique surface.

The work was performed on a digitally controlled punch press. The small, lozenge-shaped bumps go both inward and outward of the surface plain.

FIGURE 6.6 Black zinc custom perforated and bumped surface. Designed by Helix Architecture.

FIGURE 6.7 Perforated zinc used on the Cantina de Il Bruciato Winery, designed by Fiorenzo Valbonesi Cesena architects.

Small, tight perforations are also possible. Punch to die clearance should be 5–10%. Zinc can be perforated with as much variety as steels or aluminums.

Figure 6.7 shows walls created from perforated zinc. The upper image is of the walls seen glowing at dusk from internal illumination. This intriguing design was used on the Winery Cantina de Il Bruciato, designed by Fiorenzo Valbonesi Cesena. The lower image in Figure 6.7 shows the galvanized steel structural supports, which work well in concert with the zinc panels. Similar to copper and stainless steel, perforated zinc sheet offers the same metal at all edges and surfaces. Expectations of how the surfaces will perform in different climates are identical.

When using the natural zinc finish, the surface should be protected during fabrication, handling, and installation. The thin plastic coatings applied at the factory or after fabrication work well in providing this protection. Once the installation is complete, the plastic coating is removed and fingerprints or smudges that have gotten on the surface should be removed.

Custom perforating zinc (natural or patinated) benefits from the lack of work hardening of the zinc sheet as it is pierced. Other metals, such as stainless steel, will work harden as they are selectively pierced causing the metal to warp or bend and requiring post flattening of the sheets before moving into other forming steps. Figure 6.8 shows a prepatinated custom perforated zinc panel with a border and a series of panels that are only partially perforated. This is very difficult to achieve in other metals without warpage because of the shaping than can occur.

The characteristic of plastic deformation of zinc is shown in Figure 6.9. This preweathered zinc sheet was custom bumped and then formed into panels. The zinc is 1.5 mm (0.060 in.) thick. The design called for a custom trapezoidal pattern with an interlocking panel joint. When bumped and formed, the thickness and form gives the impression of a much thicker system. There is no apparent distortion on the surface. The color tone of the preweathered zinc gives the impression of steel. As the surface slowly oxidizes, no staining will develop on the concrete below.

Punching / Perforating / Bumping   183

FIGURE 6.8 Custom-perforated Baroque™ patina.

FIGURE 6.9 Custom bumped, 1.5 mm thickness zinc panels.

Zinc can take shaping well without building up internal stresses to the point where overall geometry of the panel form is affected, as it can be with copper and stainless steel that undergo levels of work hardening as shaping is induced. Figure 6.10 shows a 2-mm-thick zinc panel used on a project in Jakarta. The panels were formed from patinated zinc sheet.

These panels were bumped to create a design in the flat surface. It is important to note that the bump design was not over the entire panel. The panels did not warp out of plane as they might have with other metals, as differential internal stresses can shape the flat plane, requiring the flat sheet

184 Chapter 6 Fabrication

FIGURE 6.10 Patinated zinc panels formed from 2 mm zinc for a project in Jakarta.

to be flattened or stress relieved. With zinc, this condition does not occur, which makes zinc sheet ideal for intricate forming.

Battery casings are deep drawn into carbide tooling and dies. The zinc self-anneals from the heat of the forming operation, which allowed these deep shapes to be formed from thin zinc sheet. There are few metals that could take this kind of forming without some additional stress relief or annealing. The alloy used for battery casings is not the same as that used on architectural and art projects, but it shows the depth and dimension the metal can be drawn.

FORMING AND BENDING

Working with zinc sheet on press brake tooling is similar to working with any of the other architectural metals in sheet or plate form. Figure 6.11 shows a zinc panel rolled and shaped to large fluted forms. Essentially, the same equipment and similar forces used to shape copper are sufficient to shape zinc sheet. That being said, there are a few idiosyncratic characteristics that zinc introduces to the fabrication processes.

The manufacture of zinc into a form that could be used in art and architecture more universally did not occur until small amounts of copper and titanium were added to the rolled-zinc processes and continuous casting process came into wide use.

Prior to the modern processes for making sheet, rolling processes created a strong mechanical bias in the thin sheet. The grain direction had a profound effect on the performance of the metal and influenced how it would be used.

Zinc has a low tolerance to temperatures outside a given range. Alloying with copper helps lower the temperature, but still zinc has limitations when it comes to cold working. This places boundaries on when and at what ambient conditions the metal is workable.

FIGURE 6.11 Zinc sheet formed and rolled into a custom shape.

The general rule is that 10°C (50°F) is the lower limit of when it is advisable to work the metal using forming and bending equipment. Below this temperature, zinc will have the potential to crack along the bend.

GRAIN DIRECTION AND ANISOTROPY

Zinc sheets used in art and architecture today have alloying elements of titanium and copper in very small amounts. This gives the zinc sheet strength without sacrificing ductility. Rolled-zinc products are continually cast into a long ribbon of metal approximately 9–15 mm thick. This ribbon then undergoes further heating and pressing to arrive at a thinner ribbon of metal, usually in

0.8–1 mm thickness. In so doing this, the manufacture has effectively created a strong, thin sheet with a directional, anisotropic behavior induced by stretching the grains along the length of the sheet. The alloying elements allow the zinc crystals to slip as they are elongated preventing tearing of the thin metal. Figure 6.12 shows the alignment of the grain in the long direction.

In the closeup image, you can recognize the small-parallel grains aligned in the direction of rolling. This grain direction is in the metal and not applied to the surface as a directional no. 4 finish is applied to stainless steel sheet in post finishing processes. They may appear similar, but they are produced differently, one being a series of parallel scratches on the surface and the other being the metal itself.

This anisotropic behavior, however, does pose fabrication concerns. Zinc may tear if formed parallel to the grain, in particular when the metal is formed cold. The anisotropic nature also makes stamping of alloyed zinc sheet difficult unless evenly heated.

All rolled sheet metals have some degree of anisotropy. Anisotropy is the behavior demonstrated by thin diaphragms where mechanical properties are different in one direction versus another. Zinc, with its more defined grain, exhibits this phenomena more than the other metals commonly in use in art and architecture. The crystal structure of zinc is a tightly packed hexagonal. This structure is responsible for the anisotropic elastic behavior where the stiffness of the material is determined by the direction of the applied load.

Zinc is the only common architectural metal with the closely packed hexagonal crystal structure. Titanium and magnesium, less commonly used in art and architecture, are other metals that possess this same crystal structure and a similar anisotropy when rolled into thin sheets. The strong

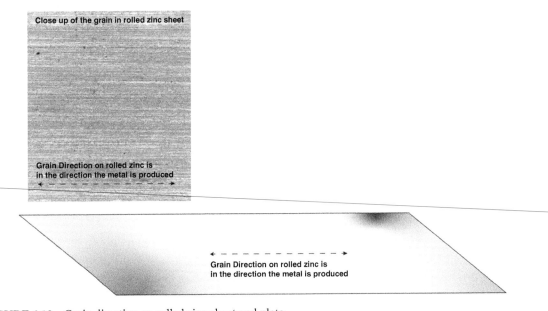

FIGURE 6.12   Grain direction on rolled zinc sheet and plate.

grain direction and anisotropic properties that rolled zinc possess require adjusting some forming processes to accommodate for these intrinsic characteristics.

TEMPERATURE EFFECT ON FORMING

One issue that sometimes arises in the field when forming or shaping rolled zinc sheet is temperature of the metal. Shaping or forming zinc sheet at temperatures below 10°C (50°F) can develop cracks along the fold line. Working with thin zinc sheet at cold temperatures will require heating zinc along the area or region where the forming is intended to occur. Most shops and plants are at temperatures above 10°C (50°F), so in plant shaping is usually not concerned with low temperatures.

In the field, if bends are to be performed by hand and the temperature of the metal is below 10°C (50°F), you can heat the metal work with a hot-air gun or gently with a propane torch. The metal should not be excessively hot, or the thin material may warp. You simply want to bring the metal temperature above 10°C (50°F) during the folding operation. Raise the temperature slowly and only long enough for the forming to occur.

BRAKE FORMING

Forming zinc across the grain direction of the sheet is similar to forming across the grain of other metals in sheet form. Zinc has a low yield strength and will shape well against the grain. There is little springback in the zinc due to the low yield stress. Springback is a dimensional change that occurs on formed parts when the pressure of the forming tool is released. For some metals with high internal stress, the bend form falls back slightly when the pressure is released. Stainless steel, for example, has a higher yield strength and higher levels of internal stresses when rolled into thin sheet. Springback is a function of a materials strength.

Forming zinc sheet with the grain calls for a more careful approach. Forming zinc with the grain should be less tight and set to a greater radius than when forming other metals of similar thickness or when forming zinc sheet against the grain. Alloyed zinc sheets lack the flexibility when the metal undergoes the needed elongation as the bend occurs. When bending with the grain, if the bend is sharp and tight, cracking may occur. The radius along the bend should be generous. Consider a radius at bends parallel to the grain of three times the thickness to avoid splitting along the grain. Figure 6.13 indicates what is meant by *with* and *against* the grain.

Folding zinc against the grain has different springback characteristics than when you shape the metal with the grain. There is more resistance; the springback component is greater than folding with the grain. However, with zinc, the yield is so low that most modern equipment will hardly notice. Hand-tooling and folding operations that might need to be performed in the field will need to contend with these differences, but they are minor.

Zinc sheet used in art and architecture is as surfacing material is often made into thin, folded, and interlocking shingles or cassettes as they are referred to in Europe. The thickness used for creating these thin panels from rolled zinc is typically 0.8 mm (0.032 inches). This thickness is sufficient

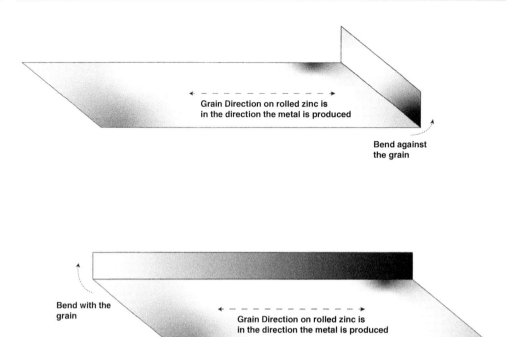

FIGURE 6.13   Bend with and against grain direction.

when the zinc is fully backed. Figure 6.14 shows a few examples of the shingle. The top two images are diamond-shaped zinc panels with the Roano™ patina. The shingles are placed on a dome shape, so each row of shingle is slightly different in size moving up the dome.

The substrate is a waterproof membrane. The backside of the zinc was painted to provide some back-side corrosion protection. Venting was not used and is the recommended procedure. The clips were stainless steel. These are well draining surfaces and any moisture that condenses or collects on the back will be removed by gravity forces.

The lower images of Figure 6.14 show similar shingles. The lower left is a preweathered gray shingle in a similar diamond pattern with an accurate fold around the corner. The lower-right image is a shingle with the Roano™ patina used as a soffit. In this instance, the panels are rectangular and applied in a running bond expression, creating a herringbone pattern at the corner intersection.

These are examples of common uses of rolled-zinc sheet for architectural metal surfaces. Zinc roofing systems and wall systems in thin sheet require a solid backing to provide support similar to thin copper sheet. It is important that the underlayment sheds moisture and does not adhere to the zinc surface. As indicated, some of the thin zinc sheet is painted on the reverse side with a thick plastisol coating. This coating allows flexibility and reduces the potential for corrosion on the reverse side to a point.

It is the susceptibility to moisture that is acidic or alkaline that poses an issue to zinc. On the reverse side, if the surface is allowed to remain moist, conditions can develop where corrosion cells

FIGURE 6.14 Zinc shingles of various shapes.

will develop. Rosin paper used on some roof applications can be deleterious to zinc because it can hold moisture against the zinc.

In North America, most underlayments for roofing are plywood, with a few instances of Douglas fir or southern pine as the supporting surface. These are acidic when moist and zinc should be separated from them. Often, OSB (oriented strand board) is substituted for plywood for economy. These contain urea-formaldehyde, isocyanate adhesives, or phenol formaldehydes, along with other substances. When OSB is wet and decays, it can damage zinc and other metals.

In Europe, the practice is to vent the surface supporting the zinc by creating an air layer separating the roof underlayment from the insulation. The air layer may be a vented space or an offset roof cavity that is vented from the eave to the ridge. The size of the space is determined by the slope of the roof. In North America, this is sometimes developed into the system that supports the zinc sheathing. Vent space can be developed with breathable subsurface material such as a nylon structural matte material. Figure 6.15 shows a couple of examples of this structural underlayment material. Often, the back side of the thin zinc sheet is painted with a thick plastisol and set over an air and moisture barrier, but it is recommended that an air space be developed to enable the reverse side to dry out in the event of condensation forming. Other systems that use breathable membranes, in particular on walls, can work as well.

Other system use formed panels made from heavier thickness zinc and set off from the surface. The thickness and form provide the strength rather than the substrate for a thinner zinc panel system. Figure 6.16 shows several examples. These are not back painted, and the thickness offers both structure to the form and further corrosion resistance.

The upper-left image in Figure 6.16 depicts custom panels manufactured from Hunter™ patinated zinc sheet of 1.5 mm. The underlayment is a breathable membrane that sheds moisture but allows air to pass. The lower left is a 2 mm thick zinc sheet formed into a lapping scale-like plate,

FIGURE 6.15　Structural matte systems composed of woven nylon.

FIGURE 6.16　Thicker zinc panel systems, folded and set off from the surface.

Roll Forming 191

FIGURE 6.17 V-cut and conventional corner folds.

and the right-hand image is a thin, custom laser cut scale-like shingle set off the underlayment with a full, several-mm-thick vented air layer.

In all these systems, the goal is to allow the back surface of the thin zinc to have access to air and allow the back surface to remain dry.

V-CUTTING

Zinc plate can be v-cut, as shown in Figure 6.17. It is not as common as other metals due to the low yield strength of the zinc and the weakness at the v-cut corner.

The method of creating corners in zinc forms is more often approached in the conventional manner where the panel ends at an edge or folds around an edge as depicted in the middle and right images on Figure 6.17.

If v-cutting is performed, consider some additional reinforcement of the corner condition. For zinc, flexure of the v-cut corner needs to be restricted. V-cutting with the grain can split if too deep. On thicker sections, a weld can be placed on the reverse side to reinforce the corner.

ROLL FORMING

Zinc can be roll formed into profile shapes with conventional roll forming equipment used on roof and wall panels. The zinc is provided in coil and fed into the roll forming station. Usually a light leveling operation is performed as the coil is fed into the first station of the roll former.

Roll forming is a common method of producing long panels of sheet metal from coils. The method involves a series of progressive rolls that gradually shape the metal ribbon into a linear form.

Each successive matching set of dies alters the shape just slightly, which causes plastic deformation along the length of the metal surface.

The calibration of the spacing of rolls and the matching dies must be set for zinc to avoid sharp bends occurring too quickly. Gradual forming of the bends will benefit the final profile development.

The roll forming dies must be made to account for the elastic behavior zinc sheet will exhibit. Much less resistance than steel, zinc will act similar to copper and aluminum.

As the metal enters the first set of forming dies, there is a stretching and slight lengthening that occurs. As the metal moves to the next station, further stretching happens, and this is accompanied by a small degree of cold working and *hardening*.

The best roll-forming operations have multiple stations spread out over several meters to reduce the stretching that must occur as the material takes shape.

Roll forming is a rapid forming process and is used to create metal siding panels and roofing panels. For zinc, the profiles are usually shallow. This is because zinc is inherently weaker than other metals, and the need for long spanning, deep, cross sections is not a design feature intended of the metal zinc.

Galvanized steel, both unpainted and painted, are commonly roll formed in any number of profiles. The steel holds the shape and provides strength. Roll-formed galvanized steel panels are used in the industrial metal building and warehouse market.

Roll-forming zinc in the field using compact, mobile systems is possible for simple roofing profiles. However, temperature behavior of zinc should be understood. If the temperature of the zinc is below 10°C (50°F) the potential for cracking at the bend is possible.

SUPERPLASTIC FORMING

Super plasticity in the context of metals, involves the ability to plastically deform and elongate without work hardening. Aluminum and zinc are two metals that can undergo superplastic forming.

Superplastic forming of zinc has not progressed into the world of architecture or art to much degree. Some of the reason lies in the lack of availability of the special alloy sheet material. The alloy is comprised of 78% zinc and 22% aluminum. Castings are available in this alloy, but sheet is not readily available. There are other aspects needed as well. Fine-grain structure is important to achieve good superplastic formed parts.

Superplastic forming of zinc can achieve elongations of as great as 1000%, similar to that achieved with thermoformed plastic and when cooled to room temperatures, the zinc retains the strength characteristics of the metal.

The superplastic process involves heating the zinc sheet up to 200 to 275°C (392°F to 527°F) along with a mold. The zinc sheet is clamped, and heated air is applied under pressure to push the zinc into the mold cavity. The zinc loses all elasticity as it reaches these temperatures, but the beauty of the process is that after heat treatments, mechanical strength can be restored once the piece cools to room temperature.

In superplastic forming, the metal sheet will expand and stretch similar to the way thermal plastic material is formed. Once the piece cools, the shape remains.

FORGING

There are several alloys specific to forging processes. Zinc is an excellent metal for cold-forging processes. Considering how zinc is used for coinage where a blank is subjected to high pressures that impart a design into the zinc. The US Mint stamps the penny out of copper-plated zinc blanks. The level of fine detail that can be achieved from forging under high pressure is one of the sought-after characteristics of the metal zinc when considered for coinage.

Forging uses specific alloys designed for hardness, strength, and elongation. The die-cast alloy, Z35841 (ZA27), is a common forge alloy as well as the custom alloys Korloy™ 3130 and Korloy™ 3330. The Korloy™ alloys have good machinability and ductility. They contain small amounts of copper and titanium, which improves creep resistance and strength. They are more commonly used in forging processes. The zinc aluminum alloy, Z35841, is used for specific industries where high impact strength is a requirement. These are not common to the art and architectural industry.

In the late 1800s, statues would be constructed by hot-stamping zinc over plaster molds, then assembled into larger sculptural forms. Zinc sheet would be imported from Europe. The zinc was soft and when heated could be easily formed. According to Carol Grissom in her excellent and well-researched book, *Zinc Sculpture in America 1850–1950*, the W.H. Mullins of Salem, Ohio, would create a plaster cast from a clay model, much the same way that some bronze sculptures are made today. However, in this case, they would make dies for hot stamping. One die, the lower die, would be made from zinc while the upper die was made of lead and attached to a drop hammer. The pieces would be formed from heated zinc sheet as the hammer stamped them over the cast zinc lower mold. Several sections would then be assembled into large sculptures. This also allowed multiple replicas of the same sculpture to be made.

Stamping of zinc to form panels is no longer a common method used in art and architecture. Stamped panels from aluminum became commonplace, and this process is still performed on occasion today. Figure 6.18 shows a stamped spandrel panel used in many state and federal buildings back in the late 1940s up until the 1960s. Many of these were stamped in zinc. Some were painted.

FIGURE 6.18   A stamped thin decorative spandrel panel.

EXTRUSION

Zinc can be extruded into various profiles; however, there is extremely limited availability. The process of extruding zinc has simply not advanced as far as the aluminum extrusion industry. The main reason lies in the strength of the metal when extruded hot. The pressures applied by the extrusion press are higher for zinc than for aluminum.

The cross-sectional dimensions are limited to small, solid, or semi-solid shapes. The wall thickness in relationship to the shape is thicker than what is often seen with aluminum. Zinc is extruded at temperatures of around 300°C and at rates of 1–20 meters per minute. The alloys that work the best are the superplastic alloy Z35841 and several custom alloys that go by the trade name Korloy™.

Finding a firm to extrude the metal will be the issue. There is no market at this time for extruding zinc, particularly with the abundance of aluminum alloy extruding companies. In most applications, aluminum can be substituted.

Stained-glass windows use edgings called *came* made from zinc. Zinc has replaced lead alloys that once were common. This seemed like an ideal material for extruding because of the small cross section. However, today it is formed into small channel-shaped zinc forms rather than extruded zinc shapes.

MACHINING

Zinc is an excellent metal to machine and achieve high accuracy in shape and surface. The speed rate should be high, using polished tools with large clearance angles and rakes to allow the cut metal to pass. Figure 6.19 shows machining small decorative coasters out of zinc plate. The lower-right image shows the plates after they have been selectively black oxidized to highlight the raised portion.

FIGURE 6.19 Machining zinc coasters for Artizan™ zinc out of 6-mm-thick zinc plate.

Zinc accepts machining similar to aluminum. Zinc plate or zinc castings can be machined using carbide tools with end clearances of 8° or less. You would first rough out the metal at fairly slow speeds and finish the surface at higher speeds. Figure 6.20 shows a large "penny" machined from 13-mm-thick zinc plate using carbide-tipped tooling. This art piece was celebrating the fact that the penny of today is actually copper-plated zinc.

SOLDERING

Soldering is a method of joining thin metals. The temperatures for soldering are low compared to brazing temperatures. Zinc is not considered for brazing but instead for soldering or welding operations. Soldering, when performed correctly, forms a metallic, waterproof seal as it joins two sections of metal together. The solder joint is weak in comparison to welding but sufficiently strong to transfer stress from thermal movements and minor loads.

Zinc sheet is especially suited for soldering with standard tin–lead solders of the 60–40 or 50–50 types. Zinc will heat up quickly and the conductivity away from the hot soldering iron is lower than that of copper. This keeps the heat of soldering contained longer at the joint. Table 6.2 shows comparative thermal conductivity of various metals used in architect.

For soldering thin zinc sheet of thicknesses no greater than 0.8 mm (0.032 in.), the lap should be approximately 10 mm (0.5 in.) and the layers should be tightly fitted. There should not be a gap greater than 0.5 mm between surfaces. The surfaces should be clean, and a flux applied before overlapping or interlocking. The flux used when soldering zinc sheet is zinc chloride. This is brushed on to the surfaces being soldered. Fluxes removed the oxides on the surface and aid in the wetting of the joint as the molten solder metal enters the joint.

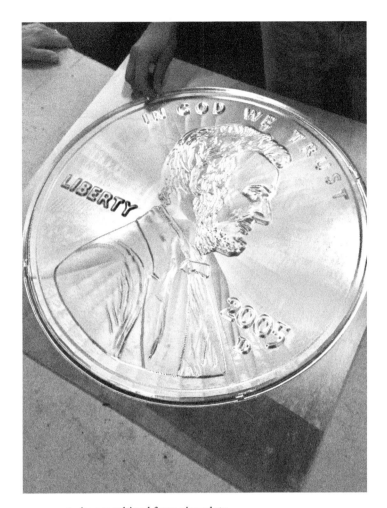

FIGURE 6.20 Large penny art piece machined from zinc plate.

TABLE 6.2 Approximate Thermal Conductivity of Various Metals Used in Art and Architecture

Metal	Thermal Conductivity W/m°K or BTU/ft h°F
Aluminum	230
Copper	400
Steel	54
Stainless steel	14
Titanium	21
Zinc	120

When soldering zinc, it is important to keep the temperature of the iron near 250°C (480°F). The heat will be transferred from the soldering iron to the zinc seam as the solder is melted and pulled into the joint by means of capillary action. Solder moves towards the heat. A large iron, preferably a wedge-tipped iron such as a hammer shape, will hold the heat.

When soldering zinc, do not apply the flame directly to the solder area, use a soldering iron with sufficient mass that will hold and concentrate the heat. Never sweat the joint with a torch. Keep the soldering iron clean and free of oxide. Consider coating the iron by pre-tinning with solder. Use only lead–tin solders. Antimony solders can corrode the zinc.

WELDING

Zinc should be welded using a shielded arc with an inert gas such as argon or helium to prevent oxidation and achieve a clean fusion of metal. The filler metal should be of the same or similar alloy. Slice off a thin, clean strip of the same metal being welded from a scrap section, if necessary. This will ensure an alloy match. The strength of the weld metal will be less than the surrounding metal.

Welding of zinc sheet can be done using gas tungsten arc welding (GTAW) techniques. This is referred to as TIG for tungsten inert gas. This process creates an electric arc between the zinc and the tungsten electrode. An inert gas, either helium or argon, are flooded over the weld area while zinc wire is manually fed into the weld zone. The welding equipment should be set to alternating current (AC) similar to welding aluminum and should be equipped with a pulse feature.

When welding cast parts, the weld zone must be free of porosity and be very clean. Porosity will cause surface blistering and weaken the weld. Zinc has a low vapor pressure so porosity will form quickly in welds. It is difficult to avoid as the metal turns into liquid.

Zinc has twice the electrical conductivity of steel but less than aluminum. Zinc also will expand more than steel and aluminum so controlling heat input is crucial. Controlling the temperature is the most difficult part. The greatest challenge with welding zinc is that once the temperature is sufficiently high to melt the zinc, you have a very tight window of time to fuse the metal before the surrounding metal collapses. With zinc, it happens quickly, due to the low melting point. Consider tack welding the parts together first then finish welding the assembly.

Figure 6.21 shows an assembly of natural zinc for a decorative feature to be used over an entryway. The tack welds are clearly visible where they have been ground back after the seam was welded by the GTAW process.

In Figure 6.21, you can also see formed and spun sections of zinc. Zinc sheet can be spun into bowl and conical forms using conventional metal-spinning equipment. Zinc is shaped on CNC (computer numeric control) metal-spinning lathes. Zinc is well suited for spinning because it does not work harden. There are various alloys of zinc well suited for spinning. These are more malleable alloy forms than the copper–titanium alloys.

There are methods of using a torch, with a very small flame to concentrate the heat at the location of the weld. This method requires a lot of finesse and careful control to prevent overheating

FIGURE 6.21 Assembled zinc ornamentation.

FIGURE 6.22 Welded v-cut zinc plate.

the joint and melting the zinc. Zinc filler rod or wire is fed into the flame at the precise point of the weld. The flame and molten metal must be carefully worked in small arcs to keep porosity in the weld low.

Figure 6.22 shows a v-cut plate of zinc, welded along the cut to add support. The metal was prepared by thoroughly cleaning the surface and the joint prior to TIG welding.

FUSION STUD WELDING

Zinc can be stud-welded similar to steel, copper, and aluminum. This is a method of attaching a stud to a metal surface by means of high-energy discharge. The capacitor discharge fuses small-diameter studs threaded and unthreaded to thin and thick metal sheet. The stud is held against the surface as a capacitor is activated to release stored energy. The energy is sufficient to melt the face of the stud and the surface of the sheet material. The stud is pushed into the molten metal, which cools and fuses the two surfaces. It happens very rapidly, and distortion is reduced.

There are a few challenges that zinc presents that differ from other metals. Zinc is a good conductor of heat and electricity. The oxide that forms on the zinc surface should be removed subsequent to the stud welding process to achieve an adequate bond. This is usually accomplished by sanding off the surface where the stud will be positioned. The oxide is less conductive and will prevent fusion, leading to poor welds.

The method used to apply studs to the zinc is the capacitor discharge stud welding (CD) process. In capacitor discharge stud welding, a specially designed stainless steel stud is used. There is a small point tip that establishes the proper gap between the stud and the workpiece. It is important that the stud sets perpendicular to the face where it is being attached. Contact is maintained by pressure using a spring in the stud gun. When the electrical current is discharged through the stud, the gaping point melts along with the adjoining faces and the spring drives the stud into the molten pool. The entire weld process occurs in milliseconds. Refer to Figure 6.23.

Stud welding is a method of attaching stiffeners to the reverse side of zinc plates. Usually, this is done in combination with high-bond double-face tape. The challenge is to limit the amount of visible show through to the face side. The addition of high-bond tape and silicone adhesives combined with the stud-welding process can often eliminate the reading of the small stud.

It is important to check the settings and the visual condition on the face of the part early in the process. Establish a protocol where the settings match with the minimum visible effect on the exposed side of the assembly. Check the tensile strength and perform a bend test by striking the fused studs with a hammer or better still use a stud pull test device to insure proper fusion. If the underside of the stud is shiny, then the surface is melting but it cooled to rapidly for fusion to occur. Additionally, there should be 360 ˆ of flash around the stud. If this is not the case, then the stud gun is not perpendicular, or something is incorrect in the stud makeup.

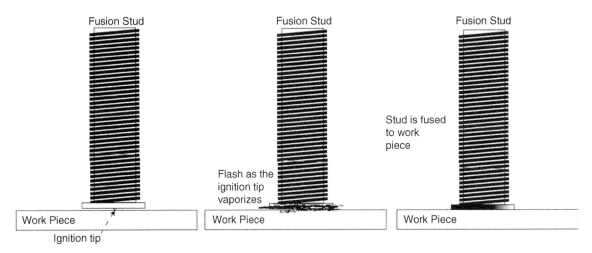

FIGURE 6.23 Stud-welding process.

With zinc, the metal will melt directly around the base of the stainless steel stud and there will be a flash around the base. Refer to Figure 6.24. The connection should be sound, but it will lack strength because the zinc will have softened and it is not a true weld but more of a braze, as the molten zinc captures the edge of the stainless steel stud. In thin zinc sheet, the stud can break off and take the zinc with it, creating a hole.

Therefore, the combined uses of double-faced, high-bond tape should be incorporated to take the loading. The stud becomes a temporary fixing and holding apparatus.

FIGURE 6.24 Fusion stud weld. Stainless steel stud to zinc sheet.

RESISTANT WELDING OF ZINC

Resistant welding, or *spot welding* as it is sometimes called, is a method of joining thin zinc sheet by passing a current of sufficient power through a small area of contact. This bonds the two metals sheets at the point of contact by fusing them together.

Spot welding uses copper electrodes, as shown in Figure 6.25. The electrodes come together and apply a small force to the top of one surface and the bottom of the other.

When the contacts touch the metal, a high current passes through the ends of the electrodes and into the zinc. This rapidly heats up the zinc and fuses the surface together at the points.

Due to the softness of zinc, it is recommended to use electrodes with rounded tips versus electrodes with more pointed or conical tips. The current should be higher than for steel.

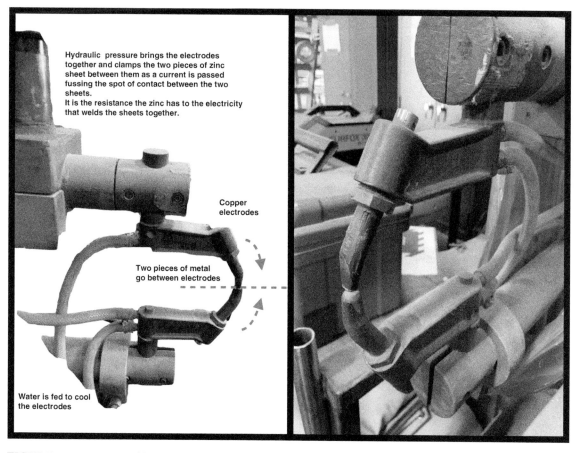

FIGURE 6.25   A spot welding setup.

EXPANSION / CONTRACTION

Zinc changes dimensions with changes in temperature at a greater rate than any of the other metals used in art and architecture. Zinc has a higher linear thermal expansion coefficient than the other metals, which means as temperatures change, so does the length and width of zinc. For rolled-zinc forms, in particular, the copper–titanium alloys form, and the anisotropy leads to significant differences in thermal-induced changes, depending on grain direction. Rolled zinc will expand and contract more in the long direction of the grain versus across the grain. For metals that are rolled into sheets or plates, this is a common characteristic, but for zinc, the differences need to be considered in design. Refer to Table 6.3 for a list of the approximate linear thermal expansion of different metals and some comparative materials.

Of the metals used in art and architecture, zinc will expand and contract more than the others. It is important to keep this in mind when designing with zinc. Zinc has a low melting point, and this correlates to a more significant coefficient of thermal expansion.[1]

For zinc, the coefficient of thermal expansion is a range between 30×10^{-6} m/m°C to approximately 35×10^{-6} m/m°C.[2] The range varies due to alloying constituents and whether with the grain or across the grain. With the grain expansion and contraction will be higher than across the grain. For most purposes in art and architecture the difference does not often play into the calculations. For zinc, the value of 33×10^{-6} m/m°C is utilized. This value means for any given length of zinc, as the temperature changes from an initial level to a different level, either warmer or colder, the length

TABLE 6.3 Coefficient of Linear Expansion of Various Materials

Material	Linear Expansion 10^{-6} m/m°C (10^{-6} in./in. °F)
Zinc	30–35
Aluminum	20–23
Copper	16–19
Stainless steel	15–17
Steel	13–15
Titanium	8–10
Glass	6–10
Rubber	150–300

---

[1] James, Spittle, Brown and Evans (2000). *A Review of Measurement Techniques for the Thermal Expansion Coefficient of Metals and Alloys at Elevated Temperatures.* Institute of Physics Publishing.
[2] Engineering ToolBox, (2008). *Linear Thermal Expansion.* [online] Available at: https://www.engineeringtoolbox.com/linear-thermal-expansion-d_1379.html.

of the zinc will change. All materials will change in size and volume as the temperature changes. The rate of change is a physical characteristic of all materials.

If, for example, a 3050 mm length element of zinc is being shaped and formed in a plant that is operating at around 10°C and the zinc part is a panel or other linear form intended to be installed in an exterior environment where the metal either heats up or cools down as the temperature or solar conditions change, the metal can be expected to change in dimension approximately as Figure 6.26 predicts. If the temperature of the metal reaches 50°C, the 3050 mm length will grow 4 mm over this temperature difference. In other words, per the graph, the length will increase to 3054 mm.

To determine how much thermal expansion to expect or to design for, use the following formula with the table of coefficient of thermal expansions for similar alloy types.

$$\Delta L = L_i \times \partial (t_f - t_i)$$

ΔL = Change in length expected
L_i = Initial length of part
∂ = Coefficient of thermal expansion
t_f = Maximum design temperature
t_i = Initial design temperature
$\Delta L = 3050 \times 0.000033\ (50 - 10)$
$\Delta L = 4.03$ mm

Designing and constructing surfaces with zinc should have the thermal expansion coefficient in mind to avoid issues of excessive movement. Designing with the blackened zinc, in particular, can offer challenges. The black color will absorb significantly more heat in the form of infrared radiation

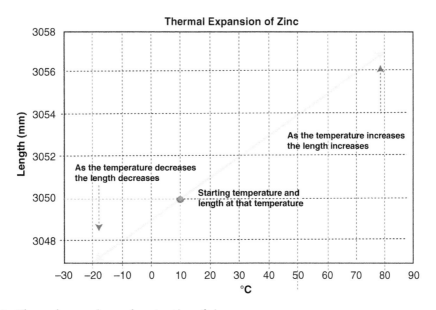

FIGURE 6.26   Thermal expansion and contraction of zinc.

than the other zinc color tones, and the temperature of the metal can rise more than the surrounding air temperature.

BOLTING AND FASTENING

There are two issues to overcome when fastening metals using screws or bolts: galvanic corrosion and pullover strength. With zinc, if the fasteners are steel, galvanized steel, or stainless steel, galvanic corrosion is usually not a concern because the zinc will act as a sacrificial material, and the ratio of areas (see Chapter 7) will have a profound effect on the speed of any reaction.

Depending on how the zinc is being held, pullover may be the challenge. *Pullover* is a term used to describe when a thin metal is subjected to a negative loading, and the metal yields around the fastener and pulls over the head.

Zinc has a greater thermal expansion than most other metals, as shown in Table 6.3. This greater thermal movement, coupled with the lower yield strength of zinc, brings pullover into the design forefront. If the design restricts the thermal movement, the hole can elongate around the fastener. This can lead to localized yielding of the zinc, allowing pullover to occur. Figure 6.27 depicts what occurs when pullover happens.

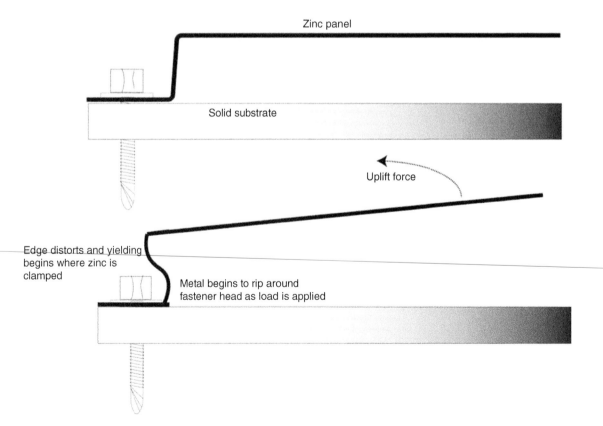

FIGURE 6.27   Pullover condition caused by thermal elongation of the fastener hole.

It is advised to allow the zinc to expand and contract away from the point of fixity and, if possible, spread the stress out sufficiently by using cleats or clamps to distribute the load rather than piercing the metal with a fastener.

In cast parts, use through bolts where possible, rather than tapping or threading. If internal threaded connects are used, use caution in designing for torque or clamping loads. Some relaxation occurs around zinc threads over time as they are subjected to a load. This is called *cold flow,* and the clamping force can reduce. Once tightly clamped, you may return to find they have loosened.

For thick plates, the joint can be machined and through bolted to create a joint that allows some thermal movement but sufficient strength to overcome shear and tearing of the joint. Figure 6.28 shows a machined joint with a through bolt connection. The hole is sufficiently large to allow the zinc to move thermally.

FIGURE 6.28 Machined and bolted connection on a zinc plate.

THERMAL SPRAY

Thermal spray of zinc, also known by the term *metallizing,* is used to add metal to a surface for corrosion resistance or to build up the surface to make it thicker. Thermal spray takes zinc wire, rod, or powder and feeds it into an electric arc or gas flame where the metal is rapidly melted and a jet of compressed air sprays the molten metal onto a substrate. In the case of a steel or metal substrate, the surface is usually cleaned and abraded by grit blasting the surface. In this way, the molten metal adheres and keys into the surface as it solidifies.

Thick coatings of zinc can be applied in this fashion. Coatings as thick as 180 µm (3 mils) can be applied using thermal spray techniques. These coatings are porous and do not provide mechanical properties of any real measure.

An advantage with zinc is its softness. Thermal spraying zinc has been used to build up a surface. It is easier to grind and smooth out than other metals. The joints can be masked so the zinc is applied only to selective areas, then ground down as needed.

Thermal spraying zinc can be applied in the plant or in the field. It is an incredibly loud process so health and safety procedures must be practiced to protect the personnel in and around the operation. The jet of compressed gas exits the nozzle at tremendous speeds and noise in excess of 100 dB is emitted as the zinc atomizes and hits the metal being coated. Additionally, the fumes from zinc are hazardous, so both the operator and any people in the area need to be protected.

HOT-DIPPED GALVANIZING

Fabrications made from hot-dipped galvanized steel are often the ancillary components involved with the support of zinc surfaces. There are designs, however, that wish to capture an industrial appeal and seek the economy and strength a steel coated with zinc can provide. Inexpensive corrugated galvanized sheets are an example of using an industrial material for architectural closure. The flexibility of galvanized steel allows a designer to create features and detail with an inexpensive medium. Figure 6.29 is an image of a newly installed horizontal corrugated galvanized steel sheet abutting a vertical galvanized corner. The corner becomes an integral part of the design rather than a "band-aid" cover to conceal inaccuracies and lack of finesse.

One of the more interesting uses of hot-dipped galvanized forms is creating with woven mesh. In this case the mesh is formed, degreased, and dipped into molten zinc. The effect is like freezing movement. Figure 6.30 are images of the initial tests in the lower image and a form designed by Malcolm Holtzman in the upper image.

The woven-steel mesh was shaped to look like draperies and fixed to steel frames. Each piece was unique. The challenge comes when immersing a shape made from woven mesh into the molten zinc. The zinc adheres to the mesh, but this adds significant weight to the mesh, so as it is removed from the tank, it has a tendency to change shape.

FIGURE 6.29 Corner detail of a horizontal corrugated galvanized panel.

FIGURE 6.30 Hot-dipped galvanized steel fabric.

Another challenge is to thoroughly clean the mesh, which occurs in pickling tanks and degreasing tanks at the galvanizing facility. There is a tendency of the mesh to hold substances via capillary attraction into the spaces created by the weave. This can cause delamination or flaking of the zinc.

The result here was a silvery gray, "drapery," that wraps the theater of Center for The Arts at New Mexico State University. Figure 6.31 shows the work as it is being installed both inside and outside the building. This hot-dipped galvanized steel surface provides strength as well as durability.

FIGURE 6.31   "Frozen drapery" made from hot-dipped galvanized steel mesh and frames. Designed by Malcolm Holtzman for the Center for The Arts at New Mexico State University.

CASTING

Zinc is an excellent metal to consider when casting small intricate forms. The low melting temperature requires less energy than other metals. This characteristic also enables the metal to be cast into low temperature molds that allow multiple reuse. The lower casting temperature can be achieved with less rigorous equipment such as induction heating sources or any number of conventional casting equipment. More extensive casting techniques such as slush casting, centrifugal casting, and pressure die casting are common industrial casting processes that utilize the low temperature nature of zinc. Add to this, zinc has excellent fluidity, allowing good detail to be obtained in the casting. Figure 6.32 shows the level of detail that can be achieved with zinc castings.

In the art world, cast zinc was extensively practiced for small monuments and sculpture in the middle to late 1800s and early 1900s in both Europe and later in the United States. Initially, the foundries in France and Germany were prolific in the production of zinc cast statues. Plaster molds were created, and multiple statues were produced, many with incredible detail. Moritz J. Seelig immigrated from Germany and opened the first zinc casting foundry in Brooklyn, New York in 1851. The M.J. Seelig & Company produced numerous statues and low relief panels from zinc. He would import neoclassical works from Europe and copy them in sand molds.

These statues were cast in sand or slush cast into plaster molds where the molten zinc is poured into the mold briefly to coat the surface then the excess would be poured back out leaving a layer of thin zinc coating the mold surface.

FIGURE 6.32   Intricate detail achieved with cast zinc.

The Monumental Bronze Company made and marketed large assembled zinc sculptures and called it *white bronze*. These would be cast into sections and welded together, then lightly blasted to give them the look of stone.

There are a number of common techniques used to cast zinc. Few today are used in art and architecture. Table 6.4 describes several of these techniques. However, the ease and ability of casting coupled with the new developments in patinated zinc possibly will bring back some of the glory zinc castings once demonstrated in the art world.

TABLE 6.4 Casting Processes Commonly Used with Zinc

Casting Method	Advantages	Limitations	Cost
Die casting	Smooth surface with good detail. Little post cleanup needed. Reuse of die.	Size limits. Limited to small parts. Specialized equipment. Metal is forced under pressure into a metal die.	Design cost high with initial first article cost high. Low cleanup cost.
Slush casting	Quick, repeatable. Takes advantage of low melting point.	Size. You must be able to pour out excess metal. The weight of the metal must be taken into consideration.	Making the initial dies is expensive. Subsequent castings are not.
Permanent mold casting	Smooth surface with good detail. Little post cleanup needed. Die is reusable.	Limits in shape and size. Larger forms than die casting. Gravity cast methods used.	Initial mold cost is high. Dies can be metal or graphite. Little waste generated.
Sand casting	Simple and straightforward gravity cast process. Large castings can be produced.	Rougher surface. Clean up and touchup of surface usually required. Detail less defined.	Low cost process. Low die cost. Higher post cleanup cost.
Plaster mold casting	Smooth surface finish with intricate detailing possible.	Low volume. Mold takes time to create.	Higher cost of mold manufacture.

DIE CASTING

In the die-casting process, molten zinc is pumped under pressure into a closed steel die. The zinc quickly solidifies, and the metal die is opened. The die-cast zinc part is ejected from the die and the process begins anew. This is a production process used to make small, smooth detailed parts rapidly. You would not consider this process unless you plan on producing thousands of parts a month.

Die casting minimizes or eliminates cleanup after casting. There is no oxide or sand stuck to the surface, and with proper temperature control, it reduces runner and gating cleanup in design. Trimming these from the casting is usually the greatest finishing cost in die casting.

Die-cast parts can be left as is or tumble finished, vibratory finished, or shot peened to create a consistent surface texture.

Zinc die casting is used in the architectural and ornamental hardware industry and smaller cast elements. Often, they are plated with copper or other metals. In North America, the majority of zinc die-cast elements use one of the following alloys:

Z33520 – also known as Zamak 3, Mazak 3, or AG40A
Z33523 – also known as Zamak 7, AG40B
Z35541 – also known as Zamak 2, AC43A

These alloys were created specifically for casting back in the late 1920s by the New Jersey Zinc Company. Zamak is an acronym for the German words for the elements used in producing the alloy: zinc, aluminum, magnesium, and copper (*kupfer*). Zamak 3, UNS Z33520, is the most commonly used of the alloys in making die castings.

There are other alloys specifically formulated for casting in all the gravity cast methods. These are commonly known as the ZA alloys due to the levels of aluminum. Chapter 2, Alloys, describe the makeup of the ZA alloys.

ZA8	Z35636	8–9% aluminum
ZA12	Z35631	10–12% aluminum
ZA27	Z35841	25–28% aluminum

These alloys offer good strength and wear resistance as compared to other zinc cast alloys.

SLUSH CASTING

Slush casting is a casting method that makes use of the low melting point of zinc. Slush casting involves pouring molten zinc into a mold, then quickly pouring excess metal out while leaving a thin layer of solidified metal on the mold walls.

Slush casting makes hollow castings with thin walls. It can make cast plates for further joining, as shown in Figure 6.33. This is a slush cast statue made of several castings, joined by welding the castings. These sculptures were common in the late 1800s and early 1900s. They could be replicated several times once the molds were created.

FIGURE 6.33 Slush cast Statue of Liberty. Several plates were cast and joined by welding.

Slush casting is not common today in art and architecture. This technique is still used to make small items and ornamentation. The difficulty is the weight of the zinc. Handling the molten metal in and out of a large die requires equipment capable of managing the weight.

PERMANENT MOLD CASTING

The permanent molds used for zinc are machined from steel or graphite blocks, usually two matching parts. The zinc is poured into the cavity within the mold, and once cooled, the mold is split apart to remove the part.

Due to the high cost of the mold, this process is usually designed for the production of multiple parts. The surface is very good and little post cleanup is needed. The size that can be produced is larger than the small die-cast parts.

SAND CASTING

Sand casting zinc is similar to sand casting other metals. A mold is made in refractory sand and the molten zinc is gravity fed into the mold. The mold is made from a pattern, usually wood or metal, that is set into the moist sand. The sand is compacted around the pattern and dried. Once the pattern is removed, a cavity remains.

The size is limited only by the facility. The surface is not as smooth as the permanent mold or the die-cast methods of casting, but the process is lower in cost and works very well with lower-volume production.

PLASTER MOLD CASTING

This is similar to the lost wax technique. Essentially, a pattern is produced in wax, foam, or other material and covered with a slurry of a special plaster material and allowed to dry. The pattern is removed, and the mold is baked. Once the mold is hardened, zinc can be poured into the mold. The zinc surface will be smooth and can have very good detail produced. Reference Figure 6.34.

Slush casting, where the zinc is poured into the mold and then poured out, is performed in these plaster molds. As the molten zinc is poured into the plaster mold, some of it solidifies on the plaster surface. Thickness can be developed by repeatedly pouring more molten metal into the mold.

SPIN CASTING

Spin casting is performed with zinc alloys. Small parts in diverse industries are created by spin-casting processes. Molds can be made of silicone rubber. The parts are quickly cooled. It is mentioned here as a process used on zinc but not necessarily a process used in art and architecture.

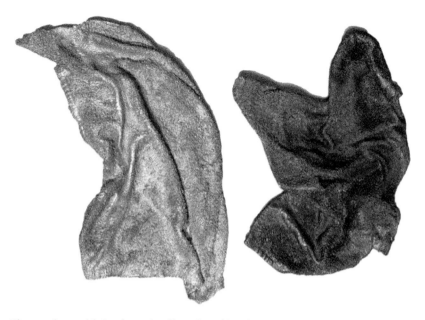

FIGURE 6.34 Zinc castings with intricate detail produced by plaster slurry techniques. "Throwing in the Towel," by Michael Wickerson.

When casting with zinc, the weight per unit area is high. Zinc is a heavy metal as compared to aluminum. Zinc is a dense metal, nearly as dense as iron. Table 6.5 shows the weights of the various metals to achieve a slab 6.35 mm (0.25 in.) thick.

Zinc is slightly lighter and less dense than stainless steel, steel, or weathering steel. Because zinc is somewhat weaker mechanically than these metals, thickness of the material may need to be increased for some applications.

TABLE 6.5 Comparative Weights of a Thick Casting

Metal	Density $1000 \times Kg/m^3$	Thickness	Kg/m^2	lb/ft^2
Stainless steel	7.972	6.35 mm (0.25 in.)	50.38	10.40
Aluminum	2.713	6.35 mm (0.25 in.)	17.08	3.53
Iron	7.861	6.35 mm (0.25 in.)	49.52	10.22
Weathering steel	7.972	6.35 mm (0.25 in.)	50.38	10.40
Copper	8.968	6.35 mm (0.25 in.)	56.50	11.66
Zinc	7.141	6.35 mm (0.25 in.)	45.00	9.29

TABLE 6.6 Mechanical Properties of Zinc

Property	Imperial	Metric
Coefficient of thermal expansion	15.4 μin./in.°F	27.7 μm/m°K
Electrical conductivity	26% IIACS	
Thermal conductivity	60.5 BTU/ft·hr·°F	105 w/m°k
Shear strength	24–28 ksi	165–193 MPa
Tensile strength	21–28 ksi	145–193 MPa
Hardness R 15t	50–68	200–400 MPa
% elongation	30–45%	
Modulus of elasticity	13.05×10^6 psi	96.5 GPa
Elastic limit	11–23 ksi	75–160 MPa

Table 6.6 shows some approximate mechanical properties with zinc (example only; check actual material). The appendix has some comparative data to other common metals used in art and architecture.

(Note that for zinc, there is a wide range in many of the mechanical properties due to the anisotropic nature of the metal.)

CHAPTER 7

Corrosion

Too long a sacrifice can make a stone of the heart.
O when may it suffice?

Source: William Butler Yates, Easter, 1916.

INTRODUCTION

Zinc has been around for the better part of two centuries; however, its current architectural alloying configuration is but a few decades old. It was not until the 1960s, producers of the architectural alloys incorporated techniques of continuous casting and alloying with copper and titanium. These two concepts advanced zinc from being simply a sacrificial coating of steel to a surfacing material to compete with copper, stainless steel, and aluminum.

Much of the research on the corrosion resistance of zinc has centered on its use as a coating for steel. We tend to think about the sacrificial behavior of zinc with its unique, natural tendency to protect other metals while it decays. But the real crux of zinc is that it acts as a very stable, unreactive barrier tightly adhered to the steel substrate. This stable barrier afforded by this thin zinc layer is why zinc sheet performs very well as a thin cladding material.

In a nonmechanical perspective, for every metal there is a stress or tendency to corrode that is induced by the environment the metal finds itself exposed to. There are distinctive climate zones, cold, moderate, dry, coastal, warm, humid that on a macroscale work to change the surface of the metal. There are also localized "climate" zones induced by surface geometry, sheltered surfaces, sloping surfaces, proximity to other metals and in northern climates the propensity of deicing salts used on our sidewalks and roadways that work on a microscale.

In the following pages, these will be discussed, as well as their effect on the metal zinc and hopefully it will provide some insight into how the revival of this old metal is responding to the environments of today.

ZINC AS A PROTECTIVE COATING

The majority of zinc used today is as a sacrificial metal or, more specifically, a protective coating for steel. Zinc forms a barrier that protects the underlying steel by providing a tight, metallurgic bond with the steel. There are several alloying layers that develop between the steel inner core and the nearly pure zinc outer shell. Additionally, as the zinc surface combines with negatively charged ions of hydroxides and carbon dioxide in the atmosphere, zinc effectively forms a stable, nonreactive layer of protection on the outside of this tight jacket. The other, unique aspect of zinc is its ability to extend protection in the event of a breach in the coating or to the uncoated edge on the occurrence of a piercing or shearing of the metal. Few materials have this unique attribute of extending sacrificial protection over a distance of separation.

When exposed to air, the zinc forms a tight oxide film. Further exposure to moisture in the air and the surface creates zinc hydroxide. Zinc hydroxide is the thick, amorphous, whitish corrosion product often seen on zinc when moisture is allowed to remain on the surface or between layers of zinc. Normally, however, it is a thin, light haze on the surface. As exposure continues, the zinc hydroxide combines with carbon dioxide and forms a thin coating of insoluble zinc carbonate in combination with the hydroxide. The chemical reaction occurs in three stages, often happening at different rates over the surface of zinc exposed to a clean atmosphere:

$$Zn + O \rightarrow ZnO$$

$$ZnO + (OH) \rightarrow Zn(OH)_2$$

$$Zn(OH)_2 + CO_2 \rightarrow 2ZnCO_3 \cdot Zn(OH)_2$$

Zinc is a bit of an enigma when considering corrosion resistance. Most designers and engineers think of zinc as being reactive due to its position on the galvanic scale, also known as the *electromotive series of metals*. Zinc falls at the far end, down with the "least noble" metals, usually accompanied by magnesium. The galvanic series is a list showing the relative position of metals from an electrochemical standpoint when exposed in a given environment, usually seawater. Another way of looking at this is the electromotive chart of metals. The electromotive force or the theoretical electrochemical potential of zinc places it on the anodic or electrically negative end of the scale. See Table 7.1.

This implies that zinc, when coupled with nearly all other metals, will corrodes first when conditions are such that a galvanic coupling occurs. Galvanizing of steel makes use of this electropotential between zinc and iron. When a galvanized steel article is exposed to a corrosive environment, zinc sheds its electrons into the steel, which makes the zinc ion positive. This, in turn, causes the zinc to combine with the negative hydroxyl ion set up in the electrolyte.

According to the electropotential relationship, the less noble metal when coupled with another metal will protect the more noble metal by means of the concept known as cathodic control protection. In the case of zinc on steel, a cathodic overprotection is in place, which lowers the corrosive tendency of the steel by passing a cathodic current through the steel. This changes the potential of the steel, making it inert in relation to the zinc.

TABLE 7.1 Electro-Potential Relationship of Metals in Seawater

Electrical potential of various metals in flowing seawater		
	Voltage range	
ANODIC POLARITY		
The more active end of the	−1.06 to −1.67	Magnesium
Scale – Least-noble metals	−1.00 to −1.07	Zinc
	−0.76 to −0.99	Aluminum alloys
	−0.58 to −0.71	Steel, iron, cast iron
	−0.35 to −0.57	S30400 stainless steels (active)
	−0.31 to −0.42	Aluminum bronze
	−0.31 to −0.41	Copper, brass
	−0.31 to −0.34	Tin
	−0.29 to −0.37	50/50 lead–tin solder
	−0.24 to −0.31	Nickel silver
	−0.17 to −0.27	Lead
	−0.09 to −0.15	Silver
	−0.05 to −0.13	S30400 stainless steels (passive)
	0.00 to −0.10	S31600 stainless steels (passive)
	0.04 to −0.12	Titanium
	0.20 to 0.07	Platinum
The more noble end of the scale	0.20 to 0.07	Gold
	0.36 to 0.19	Graphite, carbon
CATHODIC POLARITY		

In the event of a breach in the zinc coating, as shown in Figure 7.1, ions of zinc go into solution and combine with the oxygen and hydrogen ions. The exposed steel is polarized by the surrounding zinc and the zinc oxidizes. The zinc hydroxide that forms redeposits along the exposed steel. In this way, the steel does not oxidize and is protected by the corroding zinc.

In many ways, this is one of the most valuable aspects of zinc. Other metals will act as barrier coatings, but they do not offer the ability to extend galvanic protection in the way zinc does. For instance, aluminum coatings or tin coatings on steel produce excellent barrier coatings but lack the ability to extend galvanic protection to exposed steel in the event of a scratch. Zinc, on the other hand, extends protection to the steel even in corrosive environments.

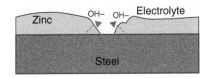

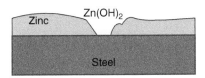

FIGURE 7.1   Diagram of zinc coating on steel.

Zinc coatings on steel also protect the steel by forming a barrier between the steel and the environment. Zinc additionally forms zinc corrosion particles on the outside surface, providing an additional protection by slowing down the corrosion of the zinc.

GALVANIZED STEEL

The zinc applied in the galvanizing method, creates a very adherent metallic jacket over the steel, effectively isolating the steel from corrosive environments. This jacket of metal forms in blended layers of zinc–steel alloy at the interface of the metals to nearly pure zinc at the surface.

With zinc coatings on steel, the zinc, acting as the anode in the coupled relationship with steel, will be attacked by corrosive environments and thus will ensures that the steel is the cathode in the relationship, remaining untouched and passive.

For the most part, the zinc that makes up the outer layer of a galvanized steel article will respond in a corrosion context similar to zinc. It will develop the zinc hydroxide and then convert, if the atmosphere is right, to a zinc carbonate.

Initially, the zinc being exposed to the atmosphere will develop a layer of protective corrosion products with a high resistivity to change. These products slow down further corrosion as the galvanized article is further exposed to the environment.

The level of protection provided by zinc coatings over steel is directly correlated to the thickness of the zinc coating and the environment the coated article is exposed in.

The thicker the zinc coating, the more protective it is to the underlying steel and the more corrosion resistant the surface is.

Galvanized coatings of zinc are resistant to normal atmospheric exposures. In some rural areas, the exposed surface of a galvanized panel element made from sheet material can darken, even as

Zinc Alloy Coatings on Steel

TABLE 7.2 Various Other Coatings Similar to Galvanized

Name	Major alloying constituent	Comment	Use
Galvalume	Zinc and aluminum	Considered twice as corrosion resistant as galvanized	Flat sheet and coil
Galfan	Zinc, aluminum, and small amount of mischmetal[a] *Mischmetal* means "mixed metal," taken from the German term, *mischmetall*. Stands for a group of rare-earth metals, cerium, lanthanum and neodymium along with traces of the other rare-earth metals.	Considered twice as corrosion resistant as galvanized	Flat sheet and coil
Zinc-alum-magnesium	Zinc, aluminum, and magnesium	Claims to be several times more corrosion resistant as galvanized	Several trade names. Flat sheet and coil

the panels on either side remain lighter color. This is a superficial condition that will eventually even out.

Temperature has no effect on the performance of the galvanized steel in most normal exposures. However, the time of wetness of a surface can lead to the development of zinc hydroxide on edges or sheltered zones.

Hot waters, waters above 65°C can cause a reversal of the sacrificial relationship of zinc and steel. At these temperatures, in the presence of moisture, the zinc becomes passive in relationship to the steel. Exposed steel will corrode, and the surrounding zinc will remain untouched.

Other zinc coatings are also in common use on steel. These are binary alloys of zinc and aluminum or ternary coatings of zinc, aluminum, and magnesium. There are also coatings that involve rare-earth metals along with zinc and aluminum. Table 7.2 lists a few of these and the common name for them.

ZINC ALLOY COATINGS ON STEEL

There are several other coatings that work similar to pure zinc in the hot-dipped galvanizing process. These are alloys of zinc coating by hot dipping steel. They all contain aluminum to some degree. It has been known that aluminum with zinc improves the corrosion resistance of the coating as far as the barrier aspect. These coatings reduce the cathodic protection afforded by the pure zinc. One is galvalume, develop originally by Bethlehem Steel. Galvalume™ is a hot-dipped coating composed of 43.5% zinc and 55% aluminum and approximately 1.5% silicon. This alloy coating goes by several brand names: Aluzinc, Zincalume, and others. There is also a coating known as *galfan*, which is composed of zinc with 5% aluminum and a small trace amount of mischmetal (rare-earth). The galfan coating retains the cathodic protection of pure zinc.

There are several coatings composed of zinc, aluminum, and magnesium with varying quantities of aluminum and magnesium, depending on the mill or market approach by the company presenting them. They are characterized by aluminum between 3% and 6%, magnesium from 0.1% and 3% and go by specific trade names that tout the benefit of these alloy coatings in respect to galvanized coatings of zinc.

Aluminum does improve the corrosion resistance of the coating in the context of a barrier. It reduces the occurrence of white rust that corresponds to zinc corrosion products on galvanized steel coatings. The lifespans of the coatings are improved, and the coatings do not have to be as thick. Most all of these coatings are made for the automotive industry, which is also seeking to improve the performance of the steel used without adding weight. Some of these coatings are being used in uncoated architectural uses. They will perform adequately in mild exposures.

Keep in mind that one distinction is that these coatings are applied to sheet in a continuous coil coating application. The coatings are consistent and thin. They are not established for coating large, fabricated sections as the batch galvanizing process is capable of doing.

Benefits of Zinc as a Coating

- Air and moisture tight jacket over a steel substrate
- Metallurgical bond between the steel and zinc
- Formidable oxide layer on the surface of the zinc
- Ability to protect steel over a distance of separation

ZINC POWDER IN PAINT COATINGS

Zinc powder can be placed into organic coatings and render similar sacrificial behavior toward protecting the coated metal. Zinc-rich primers composed of organic resins such as epoxy have zinc powder mixed in. There are thick inorganic coatings that incorporate zinc powders in a silicate binder. These hold a high amount of zinc but require proper surface preparation. These coatings act in a similar fashion to galvanized coatings by creating a paint barrier and offering galvanic protection from the zinc powder in the event of a breach in the coating.

The major benefit with these paint coatings lies in the ability to field apply the coating. Preparation of the surface is crucial to ensuring proper adhesion, and they can be used in conjunction with galvanized coatings that have been damaged. They will not match in appearance and are not considered an aesthetic coating or finish coating.

SHERARDIZING

There are other techniques that use zinc on steel to provide galvanic protection. Sherardizing is a process that coats small articles with zinc. The process was developed in the early 1900s by Sherard Cowpers–Coles. The process was not put into use until the 1920s and still today is not in widespread use. Sherardizing is an improvement over the paint coatings with zinc powder. Sherardizing diffuses a layer of zinc onto the steel part. The steel article is placed in a container with fine sand and zinc powder. The mixture is heated in a container to 375°C (707°F), just below the melting point of zinc. The container is tumbled for several hours at this temperature. The result is a very uniform, soft, grainy surface with a dull gray color. Sherardizing produces a very adherent, even, but rough surface that will receive further coating well. For obvious reasons, it is not suitable for large elements. Tumbling large articles in a drum while maintaining a constant temperature makes this process suitable for small articles only. The corrosion resistance achieved is similar to that of galvanizing.

THERMAL SPRAY

Another technique involves thermal spray. The steel article is prepared, usually by blasting the surface first with corundum, steel shot, or other coarse abrasive. The zinc, in the form of wire, is fed into a special gun, where it is rapidly melted and sprayed by a jet of air onto the steel surface. The molten metal keys into the surface and solidifies forming a coarse surface coating.

The usual operation involves feeding two wires through a special gun where the wires melt by a plasma arc that develops between the wires. A jet of air sprays the now-molten metal onto the surface of the steel.

Thermal spray zinc surfaces are thick, usually around 100 µm, but there is some variation in thickness due to the way the process works. The importance is not the appearance as much as the thickness and adhesion. If the steel surface is not prepared correctly, the zinc can spall from the surface.

Thermal spray coatings offer good corrosion resistance. They are porous so it is required to seal the surface with a low-viscosity fluid that enters the porous surface and aids in smoothing out the surface. Thermal spraying can receive paint and can be applied in the field. Essentially, if the surface can be abrasive blasted, it can be thermal sprayed.

Usually, pure zinc wire is used but alloys of zinc and aluminum wire can also be thermal sprayed. Thermal spraying can also be accomplished with zinc powder fed into a gas flame spray gun. This is not as effective as the electric arc type gun use with wire. Both processes, arc and flame spray, are loud due to the nature of the jet blast used to deliver the molten zinc.

All of these processes with the exception of hot-dipped galvanized steel have limited use in art and architecture. They may be used in the structure to support architectural or artistic features, but they are rarely used as finish aesthetic features. Table 7.3 introduces some of the limitations of the various coatings.

TABLE 7.3 Methods that Utilize Zincs Protective Nature

	Cost	Corrosion protection	Application	Geometry limitation	Other
Hot-dipped galvanized	Low	Excellent	Sheet, formed parts	Large or small	Factory only
Electrogalvanized	Low	Good	Sheet, limited shapes	Limited	Factory only
Organic and inorganic coatings	Medium	Good	Sheet, formed parts	Large or small	Factory or field applied
Sherardizing	Low	Excellent	Small parts	Small parts	Factory only
Thermal spray	High	Excellent	Formed parts	Large or small	Factory or field applied

ZINC ANODES

Zinc anodes are used in water heaters to protect the internal pipes and heating units. Zinc anodes are also used in marine applications where the anode is placed near the propellers and hulls of ships and provides protection. These are often called *zincs* and sold as a protective anode for marine craft.

Zinc anodes use the attribute of the metal to extend protection to areas away from the physical metal. As the zinc is consumed by corrosion, it protects the metal surfaces nearby. No other metals can provide this unique attribute. Zinc anodes are typically zinc rods or special shaped castings.

BATTERY

Zinc's relationship on the electromotive chart makes it ideal for batteries. Those ubiquitous double-A batteries used in so many everyday items have a zinc inner shell that produces a current to operate your low-voltage flashlights (see Figure 7.2).

The interaction of zinc with other metals to create the flow of electrical current from a battery was credited to Alessandro Volta, who back in 1799 created his famous voltaic pile. The voltaic pile was an assembly of zinc and copper discs separated by cloth soaked in brine. A current was found to flow through a wire connected to each end.

The sacrificial nature of the zinc makes it anodic, more electrically negative in potential when coupled with other metals. As the zinc undergoes corrosion, there is a loss of coating that occurs. This determines the service life of the coating. Thus, the thicker the coating, the longer the service life. The time it will take to corrode is highly dependent on the time the zinc surface is wet and the type of pollutant in the environment.

For the hot-dipped galvanized steel surface, there have been numerous studies to arrive at what to expect for a given thickness and a given environment. One study[1] found the expected lifespan of

---

[1] J. Edwards, *Coating and Surface Treatment Systems for Metals* (ASM International; Finishing Publications LTD., 1997).

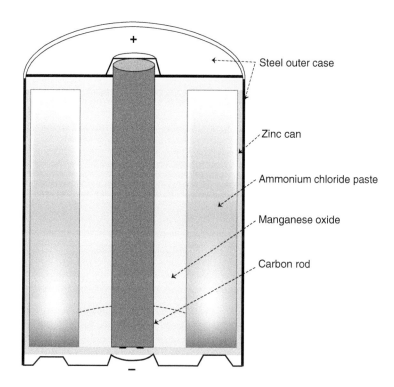

FIGURE 7.2   Inner workings of a zinc–carbon battery.

TABLE 7.4   Approximate Years of Service Before Signs of Steel Corrosion Are Visible

Thickness μm	Interior	Nonpolluted urban	Nonpolluted coastal
50	40 + years	15 years	10 years
100	40 + years	35 years	20 years
150	40 + years	40 + years	30 years

different coatings and exposures to be as shown in Table 7.4. These are batch hot-dipped galvanized steel surfaces. Coil galvanizing does not reach these thicknesses.

For galvanized coatings on steel, the useful lifespan of the coating depends on the time the surface is wet or moist and the type of pollutants in the atmosphere that effect the strength of the electrolyte. Strength of the electrolyte is related to the electrochemical nature and is reflected in the pH of the solution. Zinc is essentially consumed over time of exposure until it can no longer protect the steel and the steel corrodes. Figure 7.3 shows a corrugated galvanized wall that is corroding. From the appearance, moisture was most likely on the zinc surfaces while stored. This corrugated wall is made from thin steel sheet with a continuous hot-dipped coating of zinc. The zinc was thin

FIGURE 7.3 Corrugated galvanized steel sheet.

and easily damaged as moisture remained on the surface during a period of storage where each sheet was stacked on the other. The reverse side of the galvanized sheets would have a similar corrosion pattern.

There are three key factors that are special and unique for zinc among all the metals. These factors make zinc enormously important as a cladding material for steel and as the protector in most environments.

Key Factors Unique to Zinc
- High overpotential for hydrogen reaction in aqueous exposures
- Fast-reaction kinetics
- The formation of a tenacious oxide

The overpotential for the hydrogen reaction in electrolytes for zinc is −0.77 volts, compared to iron with an overpotential of −0.15 volts. This sets up a strong polar relationship in a solution that releases hydrogen and combines with the OH- in the solution.

This reaction happens very quickly and provides the ability to extend protection a distance away from the zinc. Aluminum forms the oxide layer instantaneous on exposure, making it passive and thus lacking the ability to protect exposed steel the same way zinc can.

The oxide, once it forms, is porous but very adherent. This corrosion product protects the underlying zinc from being attacked by other substances in the environment.

WHEN ZINC DOES NOT PROTECT STEEL

It is well understood that zinc protects steel in various exposures by sacrificial action in most natural occurring environments. Exposed edges of the steel or damages through the zinc to exposed steel will be protected by zinc's ability to extend this sacrificial action. The limit of this protection to exposed steel is approximately 3 mm of the zinc coating. This is due to fast reaction kinetics of zinc. It acts rapidly to protect the exposed steel surface or edge.

This, however, is not always the case. At temperatures above 60°C, zinc can become passive to the point that zinc and steel reverse in polarity. When this occurs, steel becomes the anode and can be sacrificial to the zinc. The steel will corrode and can fail prematurely. Galvanized steel piping carrying hot waters above 60°C will corrode the steel due to this reversal of polarity of the two metals.

At high temperatures, such as a fire, the zinc coating will perform well. If the galvanized steel is maintained at high temperatures, temperatures greater than 200°C (390°F), there is a potential for what is known as zinc peel, where the outer layer of zinc peels away from the inner, iron-steel compounds. This is because of metallurgical changes in the zinc layer interface with the iron zinc alloy layers. Zinc diffusion creates small voids that unzip the bond at the point of interface with the steel.

ZINC CORROSION

As zinc corrodes, it forms several insoluble compounds over the surface – usually in a layering effect with zinc hydroxide, zinc carbonate, then an outer layer of zinc chloride or zinc sulfide. The stability of these layers will have a direct relationship on the corrosion resistance of the zinc surface in most atmospheric exposures.

$$ZnO + 2H+ \rightarrow Zn2++H2O$$

$$Zn2++H2CO3 \rightarrow ZnCO3 + 2H+$$

The carbonate improves the passivation of the zinc surface by forming a zinc carbonate layer. In solutions where the pH is greater than 9, however, the zinc hydroxide is more stable than the zinc carbonate.

Other zinc corrosion products form, depending on the environment. These include sulfates and chlorides. Sulfates form in industrial environments over the outer layers of hydroxide and carbonates and chlorides form in marine environments.

Environment	Compound	Common name
Rural	ZnO	Zinc oxide
	$Zn(OH)_2$	Zinc hydroxide
	$2ZnCO_3 \cdot 3Zn(OH)_2$	Zinc carbonate and zinc hydroxide
Industrial and Urban	$ZnSO_4 \cdot 4Zn(OH)_2$	Zinc sulfate and zinc hydroxide
Marine / coastal	$ZnCl_2 \cdot 4Zn(OH)_2$	Zinc chloride and zinc hydroxide
	$ZnCl_2 \cdot 6Zn(OH)_2$	

When a thin layer of moisture develops on a zinc surface, the formation of the zinc carbonate will occur as carbon dioxide dissolved in the moisture joins with the zinc surface. In clean, damp air, $2ZnCO_3 \cdot 3Zn(OH)_2$, the carbonate and hydroxide develops. In dry, arid climates, the zinc oxide forms slowly but the carbonate does not develop. Moisture is needed for the carbonate to develop.

The various zinc alloys can be expected to perform close to the same in the context of corrosion and exposure. The preweathered surfaces should perform better in most cases due to the already developing protective layer. In galvanized coatings of zinc, copper, up to 0.82%, has shown some beneficial corrosion resistance in certain atmospheric conditions.

The zinc alloys are amphoteric, which means they can be dissolved in both acids and bases. Figure 7.4 shows what to expect in metal loss when the zinc surface is exposed to waters with various pH ratings. The scale on the left is milligrams per day over a surface area of 1 decimeter by 1 decimeter.

In low pH solutions, that is acidic solutions, zinc does not form an impervious barrier of corrosion products on the surface. Thus, it continues to corrode. Strong acidic environments will rapidly dissolve zinc. On the alkaline end of the scale, beyond 12.5, zinc will also experience an acceleration in corrosion rates.

As indicated in the graph, zinc performs well in solutions where the pH is in the range 6–12.5. In this range, the formation of zinc hydroxide develops on the surface and protects the metal from further corrosion. In neutral to slightly basic pH solutions, zinc will form noncontinuous adhering

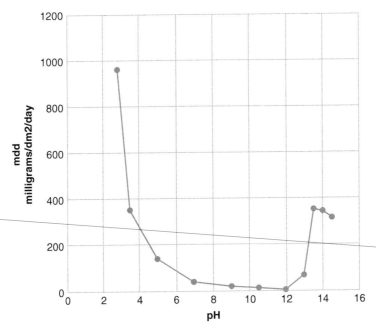

FIGURE 7.4 Loss of zinc in milligrams per square decimeters per day for a given pH.

oxides and hydroxides, commonly referred to as *white rust corrosion*. In solutions with pH from around 9 to 12, the zinc will form more continuous, adhering corrosion products that provide decent corrosion protection and passivation of the surface.

Adding phosphates and carbonates to the zinc surface will broaden the region of passivation over a greater pH range.[2] These zinc compounds on the surface improve the corrosive resistance of zinc for more severe ambient episodes.

Most of the rolled-zinc products used in art and architecture have a preweathering coating of zinc phosphate or zinc carbonate over the surface. These preweathered surfaces improve the service range and produce a less reactive surface.

Many areas across North America must contend with rainfall pH in the 4.5 –6 range. The preweathering available on the sheet products will not dissolve in these lower pH levels and have performed well as roof and wall cladding for decades.

In a general sense for zinc, when the pH is below 5, no protective corrosion products form and the zinc will dissolve. From a pH of 5 to a pH of 9, corrosion products form and inhibit further change to the surface. From around 9–12 thin, adherent corrosion products develop. The corrosion rate is very low in electrolytes that measure in this optimum pH range. As pH exceeds 13, the corrosion rate increases as zincates, forms. Zincates are ionic forms of zinc salts with various alkali salts. As these build up, they can impede the corrosion rate, and this causes the curve to drop slightly.

When preweathered, however, the boundaries are extended somewhat as the carbonate or phosphate coating developed in the factory gives a more inert, less reactive face to the environment.

INTERIOR EXPOSURES

Obviously, zinc exposed on the interior surfaces or artifacts composed of zinc, will experience a markedly different set of conditions as opposed to the exterior environment. Conditions experienced in the interior range from people touching and handling the surfaces to cleaning compounds and the general indoor environment, in particular the relative humidity.

The zinc surface and how it will perform on interior exposures is most dependent on the relative humidity. The higher the relative humidity, the more rapid the development of surface oxides or tarnish.

There have been several recent studies on the corrosion products that form on zinc surfaces exposed to various interior environments. One study involved measuring the compounds that formed on the zinc surface of artifacts stored in various interior spaces across Europe.[3] Several zinc artifacts and other artifacts made from other metals, copper, iron, nickel, and silver were exposed to typical interior atmospheres for 2 years. The surfaces were unprotected, and after the 2-year exposure, were analyzed to determine the behavior of the interior environment on the metals.

---

[2] X. G. Zhang, *Corrosion and Electrochemistry of Zinc* (Plenum Press, 1996), p. 202.
[3] C. Leygraf, I.O. Waldinger, J. Tidblad, and T. Graedel, *Atmospheric Corrosion*, 2nd Ed (Wiley, 2016), p. 93.

The surfaces all developed (with the exception of silver) a layer of corrosion products composed mainly of carboxyls. Carboxyls are a group of organic compounds composed of carbon monoxide and hydroxides along with the metals. The formula for this is COOH. These are oxides joined, in the case of zinc, to form thin layers of zinc acetate and zinc formate.

Zinc acetate	$Zn(CH_3CO_2)_2$
Zinc formate	$C_2H_2O_4Zn$

These most likely develop first as the hydroxide. The hydroxide forms rapidly on the surface of exposed zinc as moisture from the air interacts with the surface. Carbonates become absorbed onto the surface as zinc ions diffuse through the oxide and combine with the carbon dioxide in the confined interior spaces.

These compounds forming on the surface of the zinc will slow the rate of further corrosion. The surface is less reactive, and the lack of available moisture will leave these surfaces unchanged. Low relative humidity, 40% to 45%, would be sufficient to render the surface virtually unchanged.

The surface appearance is uneven tarnish. These were not preweathered, so these surfaces would readily form the oxide and hydroxide from the humidity in the air. If the surface is preweathered, these compounds would not develop.

EXTERIOR EXPOSURES

Many of the sheet alloys in use today are preweathered, which affords the surface of zinc a level of improved corrosion resistance over the unweathered surface.

Preweathering creates a carbonate or phosphate outer layer. This preweathered layer enhances the corrosion resistance of zinc by developing an inert conversion coating protecting the zinc from the environment.

Natural finish zinc will develop a layer of protective corrosion particles in time and with the right exposure, but certain environments can influence the rate of corrosion of the natural finish. Periods of moisture are the determining factor.

There was a study[4] performed in Britain back in 1964 to determine the performance of zinc as a protective material. The study set out a series of 382 zinc canisters, 31 mm in diameter and 63.5 mm tall. The canisters were mounted above ground on transmission towers distributed around the British island in various exposures for 12 to 18 months. After this, they were removed and reweighed to determine how much zinc was lost to corrosion in the various environments.

The zinc alloy used was not what is commonly used today in art or architecture and it did not possess the preweathered layer on the surface. The zinc alloy was of high purity, but the alloy had various trace elements of cadmium (maximum 0.07%), lead (1.0% max), iron, copper and tin (0.005 max).

---

[4]T. Shaw, "Corrosion Map of the British Isles," in *Atmospheric Factors Affecting the Corrosion of Engineering Metals*, ed. S. Coburn (West Conshohocken, PA: ASTM International, 1978), 204–215.

The findings of the study showed that zinc exposed to polluted urban environments fared the worst and lost as much as 4% of the total weight. Zinc in rural areas and along the coast performed very well, losing only 0.58% of their weight.

This is by no means a definitive study. Most metals, when first exposed to the atmosphere, experience the greatest rate of corrosion in the first two years. During this time, the surface interacts with the environment and builds a protective oxide layer. Initially, on some metals, zinc included, this initial oxide layer is fragile and soluble.

Other studies of the corrosion rates of metals performed in the United States show similar results. In 1955, prior to the use of the zinc–copper–titanium alloys in architecture, a study by the American Society for Testing Materials was published at the Symposium on Atmospheric Corrosion of Non-Ferrous Metals. The study examined 99.9% and 99.0% zinc exposed in an urban industrial environment, a marine environment and a rural environment. Results of the study are shown in Table 7.5.

When unweathered commercial rolled or cast zinc is exposed to the atmosphere, zinc hydroxide rapidly develops. The rate of this corrosion is slightly faster than the development of oxides on copper but slower than the oxidation that forms on steels. This corrosion happens rapidly as the surface is covered with the oxide in a matter of days.

Another atmospheric exposure test performed on unweathered zinc have shown the following reduction in thickness when the metal is exposed to various atmospheric conditions (see Table 7.6).[5]

TABLE 7.5 Rate of Corrosion in μm / Year

	Urban environment		Coastal environment		Rural environment	
	10 years	20 years	10 years	20 years	10 years	20 years
99.9% Zn	5.13	5.74	1.60	1.75	0.86	1.12
99.0% Zn	4.90	5.54	1.75	1.73	1.07	1.09

(Table created from information published on the symposium)

TABLE 7.6 Rates of Corrosion per Year

Exposure	Rate
Rural	0.2–3 μm per year
Urban	2–16 μm per year
Industrial	2–16 μm per year
Marine	0.5–8 μm per year

---

[5]Leygard, et al. 2016.

It is interesting to note that in each of these studies, zinc exposed in coastal exposures performed well, rural exposures of zinc performed very well while urban environments with their higher sulfur content, performed the least favorably. Zinc will corrode when exposed to sulfuric acid as in the case of acid rain produced from atmospheric sulfur dioxide, a pollutant more common in industrial and urban areas.

The surface corrosion that develops over time is covered with a veneer. This coating is composed of an inner layer of zinc carbonate and outer layers containing sulfides or chlorides. Galvanized coatings will develop a similar layering on exposure. You can expect the corrosion rate of zinc to be the highest in high humidity, industrial environments, where sulfur is present in the atmosphere. For zinc, and other metals used in art and architecture, sulfur dioxide is a potent adversary.

SHELTERED EXTERIOR SURFACES

Sheltered surfaces are those building surface that may be a soffit or negative sloping surface. Figure 7.5 shows examples of two surfaces that slope back at different rates. Zinc surfaces will perform differently on sheltered areas as opposed to unsheltered walls or roofs that receive abundant rain and sunshine. Rainfall tends to wash potential pollutants from the surface before they have the opportunity to transform the area into a corrosive cell. Sunlight heats up a zinc surface and reduces the time condensation remains on the surface. Both of these will have an influence on the time a surface is wet and the ability of this moisture to develop into a stronger electrolyte.

On sheltered surfaces, corrosion products can collect and there can be an acceleration of corrosion because they do not get the benefit of natural rains cleaning the surfaces.

FIGURE 7.5   Sheltered surfaces.

Sheltered Exterior Surfaces 231

FIGURE 7.6 Sheltered soffit with white zinc hydroxide forming along the interface with the concrete wall.

Figure 7.6 is a soffit clad in Roano™ patina. There is a white oxide forming back in the area where the soffit meets the concrete wall. The white is zinc hydroxide forming as these areas stay cooler and condensation forms. This increase in time of wetness, accelerates the development of the white zinc hydroxide. Eventually, this oxide development will slow down and stop as it thickens and adheres to the patinated surface.

Sheltered regions can also receive the condensation from areas above or where the waters from rainstorms tracks back by capillary action pulling with it surface contaminates that create stronger electrolytes. On the prepatinated surfaces these can be more apparent as the deposits form concentrations of white zinc hydroxide. Figure 7.7 shows the white deposits that have formed on a sheltered region of patinated Roano™ zinc. These have formed over 10 years. Further exposure should darken these deposits.

FIGURE 7.7 White deposits on sheltered area of patinated zinc.

In all exterior exposures of zinc, the first corrosion products to form are zinc hydroxide. This forms rapidly. As the surface is exposed to the atmosphere, immediately carbon dioxide is absorbed and the surface changes to a hydrated carbonate within the first days. This grows and thickens.

$$Zn(OH)_2 \rightarrow Zn_5(CO_3)_2(OH)_6$$

On the preweathered zinc, the zinc surfaces are provided with protective, factory created surfaces of zinc carbonate or zinc phosphate, so they have already advanced artificially, and this initiates a passive surface.

There is still diffusion of zinc ions through these factory-developed surfaces, but the diffusion is slow, and porosity soon fills with corrosion product to seal them. As the zinc is exposed to the urban industrial environment or the coastal environment, the availability of atmospheric substances changes these surfaces as the zinc ions diffuse out into the electrolyte and combine with substances that have been picked up from the atmosphere.

In the case of the urban industrial environment, this would be mainly sulfur. Zinc sulfate compounds form at first mixed with the carbonate and later as a more complex compound of a hydrated zinc sulfate.

$$Zn(OH)_2 \rightarrow Zn_5(CO_3)_2(OH)_6 \rightarrow Zn_4SO_4(OH)_2 \cdot 4H_2O$$

In the case of the coastal environment, chloride ions would be the more prevalent substance that would collect on the zinc surface and react to form complex zinc hydroxychloride compound. This would resemble the mineral simonkolleite an insoluble white crystalline substance.

$$Zn(OH)_2 \rightarrow Zn_5(CO_3)_2(OH)_6 \rightarrow Zn_5(OH)_8Cl_2 \cdot H_2O$$

Both of these compounds are insoluble. They will continue to protect the zinc surface in a similar fashion to the patina that develops on copper and weathering steel. Figure 7.8 is a roof of preweathered zinc installed on the coast in the Bahamas. It has developed a thin layer of the zinc hydroxychloride as indicated by the whitish substance on the roof surface.

The types of corrosion to be expected by architectural and artistic uses of zinc are listed in Table 7.7.

FIGURE 7.8 Roof exposed for 10 years in the Bahamas.

TABLE 7.7 Corrosion Categories

Corrosion	Faced by architectural exposures	Identifying symptom
Uniform corrosion	Common	Tarnish, darkening of the surface. Depends on the relative humidity of the environment.
Galvanic Corrosion or Bimetallic Corrosion	Common	Zinc is sacrificial to most all other metals. Thus, when in contact, expect the zinc to corrode. The rate of corrosion depends on several conditions.
Pitting corrosion	Not common	Not a normal occurrence in most atmospheric conditions. Copper ions will cause pitting and some alloys are more prone to developing localized pitting than others.
Dealloying	Does not occur	Even the high alloy percentage aluminum zinc alloys are not subject to de-alloying.
Stress corrosion cracking	Corrosive atmospheres	Does not generally occur because zinc is not typically used in high stress areas.
Fretting corrosion	Not common	Gouging or ripping of surface from external physical encounters is more common.
Erosion corrosion	Not common	Constant rapid flow of moisture removes oxide layer.
Intergranular corrosion	Not common	Attack along grain boundaries. More common in zinc–aluminum alloys
Corrosion fatigue	Not common	Crack in the metal across grain
Crevice corrosion	Not common	Excessive pitting under seals or washers and laps.

What determines whether the zinc alloy will experience one of these corrosion categories is dependent on three factors:

- The driving force – electron flow
- The ambient environment
- The regularity of cleaning

The driving force behind corrosion of metals is the flow of electrons. This is a natural consequence of mineral forms refined and purified into temporary existence as the pure metal form. This force is known as the electromotive force. Each oxidation and reduction reaction that a metal surface experiences has an electrical potential associated with it. If a chemical reaction is to occur spontaneously on the surface of a metal, there must be a driving force behind the movement of electrons.

This can be created by an electrolyte that develops on the surface, proximity of another metal of dis-similar electrical potential and differences in the very surface of the metal itself created by oxygen imbalances or alloying imbalances.

The ambient environment considers the relative humidity the surface is exposed to, as well as pollutants, bird waste, chloride exposure, deicing salts, even handling, and protective wraps. These will have an effect on the surface of zinc alloys, whether natural finish, patinated finishes, or preweathered surfaces. Corrosion rates of zinc are negligible if the relative humidity is below 60. Condensation coupled with inadequate ventilation will increase the corrosion rate on zinc.

For most of the metals used in art and architecture, the regularity and thoroughness of cleaning will enhance the ability of the metal surface to perform. Regular cleaning, the removal of substances, sulfate deposits, chloride salts, and other reactive substances will enable the metal surface to perform in a consistent and predictable manner.

The expectation is for the preweathered oxide or patina to be the final and constant appearance. The design intends to use the metal appearance in the original manner it was first installed. In most exposures, these surfaces change very slowly, depending on the ambient conditions, the humidity, and the pollutants in the air.

> Regular cleaning, the removal of substances, sulfate deposits, chloride salts, and other reactive substances will enable the metal surface to perform in a consistent and predictable manner.

For zinc exposed to the atmosphere, three basic salts form, depending on the ambient conditions. For many environments, you will get a combination of these salts; see Table 7.8. For all cases, where there is some moisture present, zinc hydroxide will form, and that is why all of these salts contain the hydroxide in combination with other compounds.

UNIFORM CORROSION

Uniform corrosion occurs rapidly in unweathered and uncoated zinc. Mild uniform corrosion is tarnish on the surface. Usually, tarnish manifests as fine grained in appearance, fingerprints, or simply a dulling of the overall surface.

TABLE 7.8 Various Salts That Grow on Zinc

Atmospheric conditions	Zinc salt	Compound formula	Stability
Mild / Rural	Zinc carbonate and zinc hydroxide	$2ZnCO_3 \cdot 3\,Zn(OH)_2$	Very stable
Coastal	Zinc chloride and zinc hydroxide	$ZnCl_2 \cdot 4Zn(OH)_2$	Moderate stability
Urban / Industrial	Zinc sulfate and zinc hydroxide	$ZnSO_4 \cdot 4Zn(OH)_2$	Loses stability when the pH is below 5

Uniform corrosion will occur on the surface of zinc, whether interior or exterior, and its rate of appearance is highly dependent on the relative humidity. Normally, the relative humidity needs to be 60 or greater; however, in confined areas the relative humidity can be lower. Zinc oxidation will develop on interior surfaces where dust or other particles have settled on the surface and act as hydrophilic substances. The tarnish starts at these locations and grows outward.

This type of corrosion is composed of oxides and does not harm the underlying metal. The surface can be cleaned and repolished; however, the tarnish will eventually return.

This tarnish will provide some protection to the underlying metal as it develops. On exterior surfaces the zinc oxide and hydroxide form, and from there zinc carbonate Time and exposure to clean air develops this protective layer when unweathered natural zinc is exposed to the atmosphere.

UNDERSIDE CORROSION

How the zinc surface exposed to fresh waters will perform, depends on several factors, but most importantly, the pH of the water. What other elements may be in the water, temperature of the water and oxygen content are also determining factors.

Severe uniform corrosion will occur when the zinc is exposed to waters with a pH less than 4, in particular when the moisture is maintained or regularly available.

One area where this can occur is on the underside of a zinc surface. When zinc is applied to a substrate that is not vented or when exposed to moisture, an acidic solution develops and is held in contact with the zinc surface. Often, uniform corrosion can occur. This condition can develop when certain wood products are used for support of the zinc. Condensation develops on the reverse side of the metal and the wood becomes wet releasing substances that can attack the zinc.

In the United States, plywood is often used as a subsurface support. Wet plywood can generate acetic acid, formic acid, formaldehyde, and sulfides. Other wood underlayment can be damaging to the zinc surface if they are allowed to get moist, oriented strand board, OSB, in particular can decay and develop several compounds capable of corroding metals. Newly cut oak, for example, has high levels of acetic acid. It is best to avoid wood material that when moist exudes acids and drops the pH. Try to use wood or other substances that when wet, have pH levels greater than 5. Table 7.9 lists various woods and the pH when moist.

Semi-rigid foam insulation material that contains urea can corrode zinc over time when moisture is present. Treated wood materials containing copper can pose issues and cause the degradation of zinc from the reverse side.

As discussed in Chapter 6, it is highly recommended for sheet zinc used in art and architecture on exterior exposures to be well vented on the reverse side to eliminate the accumulation of moisture. It is moisture that is needed to cause premature deterioration of the surface. Figure 7.9 shows zinc shingles in the process of being installed on a vertical surface. The left image shows the support strips, in this case made of galvanized steel. These are to allow attachment of the clips. The black material between the strips is a woven matte material made of nylon that allows air to reach the back surface of the zinc. This effectively produces a ventilation space and breaks any capillary zone that would develop if the zinc panel were flat against the wall.

TABLE 7.9 The pH of Various Woods When Wet

Wood	pH When Moist
Birch	4.8
Oak	3.9
Douglas fir	2.9
White cedar	3.4
Red cedar	3.0
Pine	>5
Spruce	>5
Scotch pine	>5

FIGURE 7.9 Zinc shingles with nylon fiber matte providing the ventilation space.

WET STORAGE STAIN

Wet storage stain is a form of uniform corrosion but has a distinctly different appearance than tarnish or light surface oxidation. Figure 7.10 shows a zinc plate with the stain. The stain is white, thick, and powdery. White corrosion of this type will form on zinc as well as galvanized steel. It occurs when moisture is allowed to remain between two sheet of zinc or when a wet environment occurs, and the air access is restricted, preventing the surface from drying or preventing carbon dioxide from accessing the surface. The stain is a mixture of several compounds, zinc carbonate, zinc hydroxide,

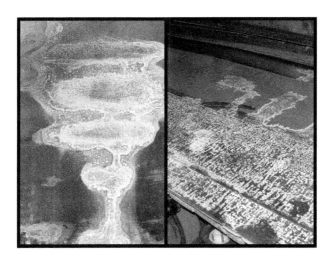

FIGURE 7.10   Wet storage stain on zinc plates and on galvanized steel.

and zinc oxide. The stain is thick, nearly 100 times more voluminous than zinc itself. Chlorides or sulfides present will accelerate the stain.

Wet storage stain consists of corrosion products, mainly zinc hydroxide. It develops when water is allowed to enter between two zinc surfaces. It will occur with galvanized steel sheets as well as zinc. Overall, it does not damage the corrosion resistance. It is very difficult to remove, however, and removal of the stain on galvanized sheet will reduce the overall thickness. It is not as common on preweathered zinc sheet, but it can develop if water remains for a significant amount of time.

The white powdery stain is zinc oxide and zinc hydroxide. To remove it one can, use abrasion but this is very difficult and generates dust. Chemically, there are some cleaning products that use gluconic acid to remove mineral deposits if they are not extensive. White storage stain does not harm the zinc. It is an aesthetic issue. Very difficult to correct on large areas.

On patinated surfaces, wet storage stain can be a major issue. Figure 7.11 shows a crate of Hunter™ zinc patina zinc with paper interleave as the protection. The crates were stored outside and allowed to become saturated with moisture. The paper absorbed the moisture and held it next to the zinc surface for a several days. The lack of air movement coupled with the moisture created the white stain along the edge. To clean these would involve removing the preweathering and the patina.

Often with metals destined for the exterior of a building, the thought is, why not store it outside since that is the metal's ultimate destiny. Too often this misguided logic is used and results in disaster. A basic understanding of the nature of a metal surface unfortunately, needs to be conveyed by the designer, the supplier and the final installer. The metal may be destined for the exterior of a building, but that environment is not the same as inside a wet crate.

Galvanic Corrosion 239

FIGURE 7.11 Crate of zinc sheets allowed to get wet and the subsequent stain on the surface.

> **Storage and Handling**
> - Keep the zinc protected from mars and scratches.
> - Store the metal in a well ventilated space.
> - Do not allow the metal to get wet.
> - If moisture enters between stored sheets, immediately dry them out.

GALVANIC CORROSION

Zinc in contact with most other metals in a humid or moist environment will be anodic. Aluminum, copper, lead, iron, nickel, and tin will all corrode zinc when in a galvanic coupling. The rate of corrosion, however, is dependent on several conditions. In this context, galvanic corrosion is different than the way zinc protects steel in the galvanizing process.

Bimetal corrosion or dis-similar metal corrosion, two other common names given to this type of electrochemical reaction involves the relationship of different metals from an electropotential position when exposed to an electrolyte. When two different metals are placed in an electrolyte, a current will flow and corrosion will occur at the anode.

Galvanic corrosion is a common corrosion concern with all metals. It is little understood by many in the art and architectural community, however. There are several factors that define the rate and intensity of the corrosion and whether corrosion will occur.

Figure 7.12 shows a typical cell and the parts needed, two different metals, an electrolyte to carry the ionic charge, and physical connection to complete the circuit.

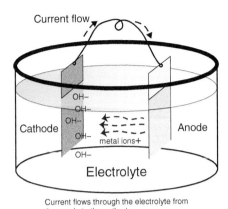

Current flows through the electrolyte from the anode to the cathode

FIGURE 7.12   Makeup of a galvanic cell.

A current develops and will flow from the anode, or more electrically negative metal to the cathode then back again through the metals where they are connected – thus, completing the circuit. The anode side oxidizes in the solution as metal ions on the anode surface join with the oxygen and hydroxide in the solution. Certain metals such as aluminum and titanium rapidly develop an oxide layer that reduces the free energy available on the surface and thus can slow the rate of corrosion.

When a galvanic couple is developed, the anodic metal is said to be sacrificial to the cathodic metal and often prevents the cathode from corroding. This is how steel with an exposed steel edge is protected by the zinc coating or a steel structure is protected by a zinc anode in proximity.

Corrosion can also occur when a copper surface is above a zinc surface. Physical joining is not necessary to cause galvanic corrosion, but an electrolyte such as dew from a roof is. Copper ions or corrosion products that become deposited on the surface can create galvanic corrosion. Figure 7.13 shows a copper roof above a corrugated galvanized wall panel. The moisture travels over the copper roof and down the wall. This exposes the zinc surface to copper ions in an electrolyte. The electrolyte is the moisture, in particular, condensation rather than rains. Rains wash the contamination away as it rushes over the surface. Condensation, on the other hand, slowly moves down the copper roof and collects free copper particles and deposits them on the zinc surface of the corrugated sheet.

In this instance, an electrolyte is present, and the zinc will corrode. No corrosion will occur to the copper from proximity to the zinc, while the zinc wall panel will undergo significant and rapid corrosion as long as the electrolyte is present. Copper ions are in the electrolyte and electrons are allowed to flow through the electrolyte. Note, condensation provides the vehicle of ion flow. If the surfaces are dry, the rate of corrosion would be very slow if at all. However, if the electrolyte is allowed to pool on the zinc surface, corrosion will be rapid.

This type of corrosion is somewhat different than true galvanic corrosion where two metals are in direct coupling. True galvanic corrosion involves the thermodynamic potential of the two metals in an electrolyte, the environment and a geometric relationship known as the "ratio of areas."

FIGURE 7.13 Image of copper roof draining over galvanized corrugated panels.

There are several processes involved in galvanic corrosion, one is the corroding or dissolution of the anodic metal in this case, zinc and the reaction is:

$$Zn \rightarrow Zn^{2+} + 2e$$

The zinc ions are dissolved into the electrolyte.

The second process that happens is the dissolved oxygen found in the electrolyte goes through a process referred to as reduction. The dissolved oxygen in the electrolyte along with the water molecule, and the electrons that are now released from the zinc react as follows:

$$O_2 + 2H_2O + 4e \rightarrow 4OH^{4-}$$

When this happens, the zinc goes into the solution as a positive ion and combines with the negative hydroxide and forms zinc hydroxide. This comes out of solution and deposits as a light corrosion stain on the surface.

This is referred to as an oxidation–reduction reaction. This is the basis for galvanic corrosion. It is an electrochemical action defined by the transfer of electrons in a chemical reaction known as oxidation and reduction. Stripping electrons from the zinc is called *oxidation*. It is not necessarily associated with the oxygen. An anodic reaction is an oxidation reaction as electrons are stripped and now available. Adding electrons, in the second equation is known as *reduction* or cathodic reaction.

The anodic metal by itself, may not react in a similar electrolyte when not coupled with the cathodic metal but when coupled, the corrosion process will continue as long as the electrolyte is present.

DETERMINING FACTORS FOR GALVANIC CORROSION

What determines whether one metal or another will corrode and the rate of the corrosion when in bimetallic coupling is related to the amount of current flowing between the metals. The oxide film that develops on metals such as titanium, aluminum and stainless steel make them passive and reduces current flow as these films lower conductivity. This, for instance, is one reason why the use of the more noble stainless steel as clips or fasteners in contact with zinc is not a concern in the environments most architectural metals find themselves. Another reason is the ratio of areas of the two dissimilar metals.

The amount of current that flows between two dissimilar metals with different electropotentials is affected by temperature, electrolyte composition, oxide and other protective barriers and by the ratio of areas. Additionally, oxygen is required. Corrosion will only occur if there is dissolved oxygen in the moisture present. Natural moisture, exposed to the air, rapidly acquires oxygen from the air so this condition is normally present. Figure 7.14 shows the categories that must be present for the development of appreciable galvanic corrosion to occur.

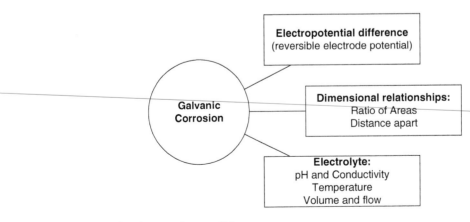

FIGURE 7.14   Galvanic corrosion conditions.

DIFFERENCE IN ELECTRO-POTENTIAL

The chart shown in Table 7.1 lists the various metals in order of their electrical potential. This is the electrical potential of two different metals exposed in seawater. There are other charts with different exposures, but seawater is often used due to its powerful electrolyte behavior. There are a number of factors that will determine whether a metal in a galvanic couple will corrode or not. The difference in the potential as shown on the chart is one, the properties of the electrolyte and the resistance of the circuit between the two metals, and how well will the current flow across this connection.

Many design professionals look at the position on the chart and arrive at the conclusion that different metals should not be joined. Whether corrosion will occur, how rapidly it will occur, and which metal will corrode requires an understanding of the principles behind this mysterious tendency of "dis-similar' metals.

The values shown for each metal listed in Table 7.1 are derived from tests of pure metals in flowing seawater, seawater being a sufficiently strong electrolyte. It could just as well be identified as water containing deicing salts, another strong electrolyte that is common in many northern climates. The electrical potential is the energy per unit area as measured in volts. It is based on the fact that two different metals will have a different potential charge when an electrolyte is present. When the charge is sufficiently different, and the metals are joined, the flow of free electrons occurs from the negative charged metal, (anode) to the more positive charge, or less negative charge, (cathode) through the electrolyte.

Many of the common architectural metals have a negative electropotential. The voltage potentials for copper, for example is negative, but zinc is more negative. The size of this difference is often thought to be the determining condition. Generally, a potential difference of 50 mV or more is considered a situation wise to avoid. But this is not the only factor. This chart is a good indicator as to which metal in a coupling would potentially experience galvanic corrosion.

GEOMETRIC RELATIONSHIP

Ratio of Areas

A very important condition of galvanic corrosion is the area relationship of the cathode and anode (see Figure 7.15). If the cathode, more positive metal area is significantly greater than the anode metal, the rate of corrosion will accelerate. If the anode area is significantly greater than the cathode, the rate of galvanic corrosion of the anode will be lessened.

For example, if a stainless steel fastener or clip is used to fix a zinc roofing panel with a cross section significantly larger than the zinc, the area of stainless steel in relationship to the area of zinc is very small. The galvanic current will be constant between the two metals but the corrosion per unit of area of the zinc is low because the current density is low. The stainless steel fastener is needed for the strength it will provide. Stainless steel is more noble than the zinc when in a passive state and the area of zinc is significant in relationship to the small area of the stainless steel fastener. In this

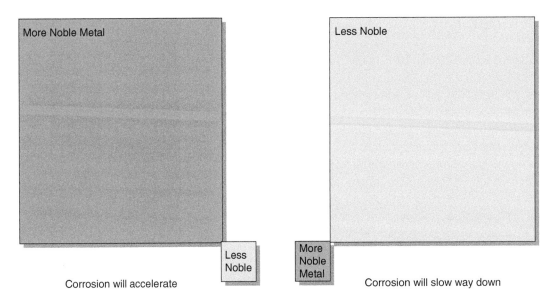

FIGURE 7.15 Ratio of areas.

case, the stainless steel will be protected by the zinc anode, and corrosion of the stainless steel, even when exposed to a strong electrolyte will be negligible. The zinc will experience very little corrosion due to its larger area in relationship to the area of stainless steel.

If, on the other hand, a zinc-coated fastener is used to secure a copper roof or flashing, the zinc coating will be severely attacked by these galvanic forces. The current will flow from the zinc to the surrounding copper the second moisture is present. The larger area of the copper will engulf the small area of zinc.

DISTANCE

Galvanic corrosion can occur between two metals in an electrolyte even when physical contact does not occur. The galvanic current that develops does decrease as distance increases, due mainly to resistance that develops within the electrolyte. Galvanic protection of steel by zinc is a form of galvanic corrosion. Zinc is able to provide protection to the edges of steel that are exposed. The dimension of the more noble metal plays a part, whereas the dimension of the zinc, the less noble metal, is not as critical because the galvanic corrosion that occurs happens at the nearest edge or surface to the noble metal.

ELECTROLYTE EFFECTS

The electrolyte itself has to have the capacity to transfer current through it. If the capacity is weak, the current will be weak. Electrolytes with a high capacity are those that have an ionic make up such as acids and alkalis, in particular those that contain chloride ions. Seawater and condensation, particularly where deicing salts are prevalent, make for strong electrolytes. Condensation from dew forming on the metal surface is more a concern than rainwaters. Condensation forms slowly and remains on the surface longer. Deionized water and distilled water are poor electrolytes. The more ions in solution the more conductive the solution is and therefore the stronger the electrolyte. Rainwater has a low capacity to carry a current unless or until the water comes in contact with pollutants. The pollutants can be airborne or on the surface of the metal and as the water sets on the surface, a more powerful electrolyte can form.

If the electrolyte is very thin, even with strong electrolytes, the corrosion will be limited and generally right at the point of contact where the two metals touch. The corrosion may still be rapid at the point of contact since here the resistance to electrical flow is the least.

The amount of oxygen and the ability of the oxygen to diffuse to the surface of the metal is another consideration in galvanic corrosion. If the electrolyte is sufficiently aerated, that is, contains a lot of oxygen, the rate of galvanic corrosion will increase.

The presence of metal ions in solution, in particularly copper ions, can rapidly corrode zinc. More noble ions in solution will develop small galvanic cells across the surface as these ions become deposited on the zinc surface.

TEMPERATURE EFFECTS

Corrosion is an electrochemical reaction that occurs more rapidly in warm regions than in cooler regions. Chemical reactions that occur when metals corrode are slowed down when the temperature is cooler. Bimetal corrosion speeds up in conditions where it is warm and humid. Chemical reactions occur more readily when heat is added. For example, conditions where deicing salts are in common use are during times of very intense cold. Chemical reactions slow way down. The deicing salts may do their job on melting ice but corrosive activity on metal surfaces does not intensify until the temperatures rise sufficiently in the spring.

For galvanic corrosion to occur, you need an electrolyte, a source for oxygen and some form of connection to allow electrons to flow from one metal to the next. Preventing any one of these and galvanic corrosion will be impeded. The diagram shown in Figure 7.16 indicates the three conditions.

When the condition exists that a connection or contact between zinc and a more noble metal cannot be avoided, it is advised to coat the more noble metal with a sound separation layer of paint or other nonconductive substance.

The reason this is critical is again, the ratio of areas. If the coating on the less noble zinc is scratched or breached in any way, this exposes a small area to bimetallic corrosion by the coupling with the larger area of the more noble metal.

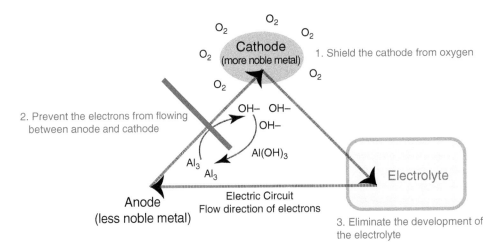

FIGURE 7.16 Galvanic circuit and means to prevent it from occurring.

This is one of Faradays laws where a given current passes between the anode and cathode in a galvanic cell at a proportional rate. If, for example, the cathode area is 10 times larger than the anodic metal area, then the current will be 10 times as great passing through the anodic metal and corrosion of the anode will be rapid.

Steps in Preventing Galvanic or Bimetallic Corrosion

- Properly design the joint to prevent water from collecting.
- Use metals close together in the galvanic series (50 mV or less).
- Respect the ratio of areas. Avoid small areas of less noble metals.
- Use a nonconductive coating to separate the metals.
- Coat the more noble metal rather than the anodic metal.
- At the junction of the two metals prevent electrolyte from collecting.
- Prevent copper salts from being deposited on less noble metals.

PITTING CORROSION

Zinc alloys will be subjected to pitting-type corrosion in hot waters or moist soils. Pitting is not normally an issue with zinc exposed to the atmosphere unless waters containing copper ions are allowed to come in contact with the surface or if very acidic waters develop on the surface.

Pitting requires the breakdown of the passive layer. For zinc, the pits often take on the shape depicted by the zinc crystalline planes. The pits can be tiny rectangular and hexagonal holes.

Allowing more noble metals to come in contact with the zinc surface while moist will cause pitting. More connected with galvanic corrosion, small copper particles from runoff will damage the passive layer and cause pitting to occur. Copper ions Cu^{2+} are more noble than zinc, and they will enhance the corrosion of the zinc surface by creating small galvanic cells where the copper ions deposit on the zinc surface.

These copper salts from the runoff of copper roofing or piping can corrode metals. What happens in that case is the salts become deposited onto adjacent or nearby metal surfaces and these cause isolated galvanic cells to develop. Usually, this is a stain that grows until the less noble metal corrodes. You can see the green stain developing over the years on limestone or concrete surfaces just below exterior copper surfaces. This same oxide, when allowed to pass onto galvanized steel surfaces or zinc surfaces, will corrode the zinc by establishing corrosion cells that eventually dissolve the zinc away.

For example, a copper gutter installed over the top of a zinc roof can corrode and eventually perforate the zinc roof if the condensate that forms on the copper drips onto the zinc. The condensate captures small amounts of copper ions, and these are deposited onto the zinc only to generate numerous small corrosion cells. Similarly, condensation from a rooftop air-handling unit, passing through a copper tube and running over a roof into a galvanized steel downpipe will cause small galvanic cells and lead to corrosion of the less noble metal. See Figure 7.17.

This is an example of where the two metals do not necessarily need to be physically connected but only need to be electrically connected. When the two metals are joined by conductive media, or in this case, ions of copper deposited on the surface, in the presence of an electrolyte, galvanic corrosion can still occur as tiny corrosion cells develop around the copper ions.

For galvanic corrosion to occur, there must be ion formation at the anode – in this case, the zinc surface. Here, oxidation is occurring, and the zinc is corroding. At the same time, there must be an acceptance of the electrons at the cathodic site, the copper.

The process will continue as the pit deepens, eventually perforating the zinc. This destructive, irreversible process can happen very rapidly.

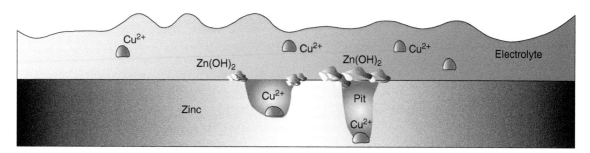

FIGURE 7.17 Pitting developing on a zinc surface.

INTERGRANULAR CORROSION

Zinc of high purity will not experience intergranular corrosion. But zinc is not often used in the unalloyed state. Zinc alloyed with aluminum is susceptible to intergranular corrosion. Intergranular corrosion attacks the grain boundaries. Intergranular corrosion appears as a series of jagged microscopic cracks running along the edges of grains or crystals. The metal will lose strength and ductility. Simple bending will fracture the metal and, when severe, the metal will begin to fall apart.

Most all die-cast zinc articles are produced by zinc–aluminum alloys making them particularly susceptible to intergranular corrosion. With the zinc–aluminum alloys where aluminum is greater than 0.03%, intergranular corrosion can occur. Zinc alloys with 0.2% aluminum are susceptible to intergranular corrosion cracking due to stress and elevated temperature. Prolonged exposure to high temperature water vapor will permanently damage zinc–aluminum alloys by causing extensive intergranular corrosion.

> What determines whether zinc will experience one of these corrosion categories is dependent on five factors:
> - The driving force – electron flow
> - The ambient environment
> - The regularity of cleaning
> - The soundness of the clear coating

STRESS CORROSION CRACKING

Zinc is not normally used in situations where it is subjected to extensive stress. Roofing and wall cladding applications may have localized stress conditions. Cold working of zinc sheet will have little effect. Stress corrosion cracking is not an issue of concern in high-purity zincs in normal atmospheric exposures.

Cracks can, however, develop from creep. Creep is a condition where the zinc would undergo plastic deformation over time under relatively light loads at room temperatures. Another name for the phenomena is cold flow, since the metal moves at room temperature rather than elevated temperature.

High-purity zinc used in casting can develop cracks from the stress of creep. Such cracks can lead to more damage, as moisture accesses the inside of the sculpture. The thickness of the zinc prevents excessive corrosion, but water can freeze and expand.

FIGURE 7.18 Cast zinc sculpture of a boy and a fish with crack.

Figure 7.18 shows a cast zinc statue. The foot of the statue has a significant crack developing. On close inspection, smaller cracks are apparent. It is difficult to say if this is solely the result of creep, because this is the place where the majority of the stress is occurring.

A condition known as *zinc pest,* from the German *Zinkpest,* is a destructive, inter crystalline corrosion that occurs on older die-cast zinc, in particular the zinc–lead alloys. This is a common yet irreversible problem, faced by collectors of small die-cast toys. Small cracks, blisters, fractures, and pitting occur, and the thin zinc becomes very brittle and friable. The parts can decay into dust.

These old die-cast toys and parts were made from castings with impurities in them. The practice was to throw various scraps and parts into a bucket for recycling. This was known as "pot metal' or "monkey metal." Parts cast from this variable purity zinc would contain lead, tin, and other metals. Once cast, they would look fine but after a few years they would suddenly start cracking and pitting.

For castings today, this problem does not exist as better controls and an understanding of the metal has reduced or eliminated this occurrence.

ZINC ARTIFACTS AND STATUES

There are several forms that corrosion takes on the surface of zinc statues exposed to the atmosphere. Similar to the green corrosion products that form on bronze sculptures and referred to as patinas, zinc also develops an enveloping oxide on the surface. This oxide incorporates other elements such as carbon, sulfur, and chlorides that are present in the urban environment. Figure 7.19 shows a steeple with dark gray and black stains that have formed over the decades of exposure in an urban environment.

Zinc surfaces can form a very adherent white crust of zinc hydroxide. This insoluble crust develops when moisture is allowed to collect and remain on the surface. It doesn't take away from the

FIGURE 7.19 Old zinc flashings and forms on a steeple in an urban environment.

appearance and can add a level of contrast. Figure 7.20 is of the same Boy and a Fish. You can see a white hydroxide that has formed on the underside or sheltered surfaces.

Zinc also forms a dark patina of carbonates and sulfides. In many of the older, urban exposures, the surface almost turns black. This mineralized surface is insoluble.

On exposed zinc, these patinas provide protection to the base metal. Removing the oxides that create the contrasting color is very difficult and will expose the base metal. This can make the contrast even greater. The exposed base metal will develop the oxide back eventually as exposure to moisture or humidity. If you remove the oxide for whatever reason you can re-age the surface by applying dilute phosphoric acid or vinegar. Wipe it on and reapply until you get the darkened surface, then thoroughly rinse the surface.

FIGURE 7.20 White zinc hydroxide on the underside of the Boy and a Fish sculpture.

Natural, unaged zinc will fingerprint readily. The acidic nature of perspiration on your hands will etch the surface and produce an adherent mark on natural zinc. Removal of these stains is difficult and requires a treatment method of mild acids to dissolve the oxide. Bar tops made of zinc require constant care to keep them looking in acceptable shape. For some aesthetics, this aged look is desirable and only a moderate cleaning to remove foreign substances is undertaken.

DEICING SALTS

So much of the architectural use of metals brings the surfaces down to the proximity of our streets and sidewalks. In the wintertime, the liberal application of deicing salts on these walkways and roadways often finds its way to the surface of metals nearby. Figure 7.21 shows a preweathered zinc surface that has had deicing salts collect and deposit on the surface.

This surface has formed a white zinc chloride with zinc hydroxide along the lower levels.

Where the flashing interrupts the panel, the partially sheltered areas do not receive the cleansing as intensely as regions above this line afforded by occasional rains.

This surface has been exposed for more than 20 years. The zinc has not deteriorated. The stain has thickened over the years, but there are no signs of pitting or premature failure of the zinc sheet.

Deicing salts are hydrophilic, meaning they attract moisture. This increases the time the metal surface is wet. Zinc and galvanized steel have shown good long-term performance results when exposed to environments that experience deicing salt spray.

FIGURE 7.21 Deicing salts reacting with the zinc and forming a stain.

CHLORIDES

A testimony to the metal zinc's corrosion resistance when confronted with chlorides is shown in Figure 7.22. This image is a vent through zinc wall surface. Inside the building is a public pool. Chlorine vapor is removed through this vent. On humid days, the vapor will combine with the water in the air or water that has condensed on the metal surface. When this occurs, hypochlorous acid forms. Hypochlorous acid is a weak acid but as it concentrates it can be very harmful to steel, wood,

FIGURE 7.22   Chloride fumes from a swimming pool.

and other materials. From the image, you can see the concentration in the vicinity of the vent is so significant that the grass around the area is dying and the concrete has a clean appearance from the constant acid wash.

The zinc appears to be holding up well, at least on the outside. The red corrosion product along the base is from the galvanized steel clips used on the base flashing. This would indicate that the zinc underside might be experiencing corrosion; otherwise, it would sacrifice to the steel. There may also be a reversal of passivity occurring making the steel be more anodic than the zinc. The whitish stain on the zinc is most likely the soluble form $ZnCl_2$. This would be indicative of the runs and collection along the edges.

FERTILIZER

Fertilizers come in a variety of types, but most common forms contain multiple nutrients such as potassium, phosphorus, and nitrogen, which are important for plant growth. These are water-soluble and often ammonium based, such as monoammonium phosphate or diammonium phosphate.

If the fertilizer is spread such that it adheres to moistened zinc, a stain or spot will form. There is minimal damage beyond appearance, as the stain tends to be lighter in color to the contrasting dark zinc surface. Figure 7.23 shows a zinc surface with staining from fertilizer.

FIGURE 7.23 Fertilizer staining on zinc.

SAPONIFICATION

Zinc oxide is amphoteric, which means it can be either basic or acidic. When the base forms it can undergo saponification with certain coatings or paints, particularly the oil-based coatings. Saponification converts the oil into soap when the zinc oxide has developed into a basic substance. This can lift the coating from the surface. An example of this is shown in Figure 7.24. In this case, a zinc coating was plasma sprayed to level out the seams in a copper alloy form. An oil sizing was used to apply gold leaf to the form.

Over time, the sizing released from the zinc portion but not from the copper alloy portion of the surface. This was due to saponification at the zinc–sizing interface only.

This condition also plays a role in the difficulty with getting paint to adhere to newly galvanized surfaces. Phosphate treatments greatly improve the receptiveness of the zinc surface to the adhesion of various paint coatings. Allowing the surface to weather for several months will roughen the surface and improve paint adhesion.

CORROSIVE SUBSTANCES IN PROXIMITY

Other substances that can have a detrimental effect on zinc, particularly in closed environments where the relative humidity is high and where the other substances can decay, are listed in Table 7.10.

Many of these substances will undergo some level of transformation in warm and moist environments. As this transformation occurs, they can exude corrosive vapors that prevent the development of the protective outer layer of oxides on the zinc surface. Usually, the surface experiences uniform

FIGURE 7.24 Gold leaf releasing from zinc surface by means of saponification.

TABLE 7.10 Various Organic Substances That Can Corrode Zinc

Substance	Effect
Phenolic resin material	Highly corrosive to zinc
Certain plastics: thermoset plastics	Corrosive to zinc
PVC, polycarbonate	Corrosive to zinc
Glass reinforced polyester	Corrosive to zinc
Epoxies – during curing	Corrosive to zinc
Rubber in proximity	Corrosive to zinc
Oil paints	Mildly corrosive to zinc
Oak, beech, chestnut, fir	Corrosive to zinc

corrosion but sometimes, due to localized humidity or moisture, the attack on the zinc can generate pitting.

Copper and water from copper flowing onto zinc can damage the zinc, as shown in Figure 7.13. Similarly, copper should not be used as a wood preservative.

The fire-retardant wood materials that use sodium silicate or sodium phosphate will not have a detrimental effect on zinc.

CHAPTER 8

Maintaining the Zinc Surface

I will not let anyone walk through my mind with dirty feet.

Source: Mahatma Gandhi

INTRODUCTION

The use of metals in art and architecture has expanded over the last century. Metals offer a defining clarity of the form of a design and at the same time provide an endless ability to shape and configure. No other material available to humankind can be used for centuries, removed, and then be melted down and reshaped to stand for centuries more. Natural zinc is one of those materials. Once we have removed it from the earth, it can remain a material of utility forever.

The useful service life of a material depends on the environment that it must interact in each day and how the material surface is maintained in the environment. With metals, they want to change. There is a drive to combine with various substances, and if left to their own inherent nature, the surfaces will change. Moisture is the catalyst for many of these changes but the metal itself, seeks out oxygen and carbon dioxide.

Zinc has this constant desire to change. It will tarnish rapidly, a matter of days, as its surface absorbs atmospheric moisture and oxygen. On natural zinc this proceeds for a few years, rapidly at first, and much of the change occurs in the first two years of exposure. Preweathered and prepatinated surfaces of zinc have already advanced in a predetermined fashion, so change for these surfaces are slower but they still change.

For so many metals used in art and architecture, the initial installation is the beginning of the exposure period and the metal surface is left to interact with the environment. The occasional rain is the limit of the maintenance. For many environments, this is all that is needed along with an understanding that the surface will adapt and reaction and slowly change.

This is often the case for zinc as well. It is the special conditions introduced to the natural environment that can lead to frustration and changes that are not aesthetically pleasing. Dents, scratches, and mars are introduced to the surface by lack of care or mishap while deicing salts are introduced by chance encounter.

Most zinc surfaces designed for our buildings and for art are intended to remain aesthetically appealing. A shabby façade or art piece is often a reflection on artistic or intellectual awareness by those that use or venture near the form.

This chapter will address a few of the conditions that often arise when working with zinc and offer suggestions to prevent them from occurring in the first place and remedy the situations when they do arise.

ZINC SURFACES

New zinc surfaces want to combine with oxygen and other substances once introduced. The shiny reflective new zinc surface will develop a tarnish if exposed to moderate levels of relative humidity. These fresh new zinc surfaces will readily fingerprint and etch if they come in contact with acid or alkaline fluids. Fingerprints may not be readily visible, but over time on smooth surfaces they appear as dark smudges.

Zinc was used extensively in France as bar tops in pubs as well as rooftops across Paris. The French slang term for a drinking establishment, *le zinc*, is a reflection of the past use of the metal as a bar top. Today you can still find a few of these special and beautiful surfaces as they have taken on the character of all the people that have spilled a bit of wine or placed a cool pint of beer. The visible remnants of conversations past.

There are several choices a designer has when considering zinc as a surface material. These are listed in Figure 8.1

Each of these surfaces must be expected to interact with the environment and provide long term service with minimal maintenance.

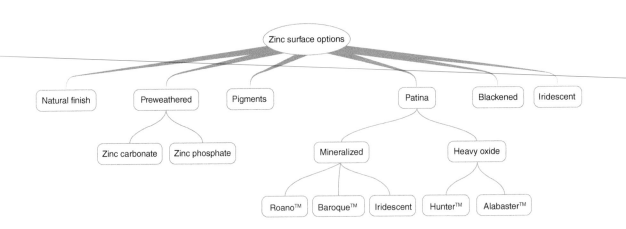

FIGURE 8.1 Choices to the designer.

The metal zinc in the natural form, whether the commercial pure material or galvanized, will change in character as moisture, handprints, and cleaning fluid interact. When preweathered or prepatinated, the surfaces are far less reactive. Fingerprints and smudges readily wipe off the surface.

The preweathering surface is produced only on thin sheet material by the factory supplying the metal. You can preweather castings and plate, but these are done on a per piece basis by postmanufacturing processes performed by companies expert in the oxidation of metals. On sheet 2 mm and less in thickness, this preweathering is done on coils in a rapid process by the mill producer of zinc.

On thin sheet or coil material, the typical surface is distributed with a preweathering treatment. This preaging or preweathering is either zinc phosphate treatment or a light zinc carbonate treatment to the zinc surface. This darkens the metal to a matte blue-gray sometimes with a light olive tint. The point of these preweathering treatments is to impart a nonreactive, even color tone to thin sheet.

Newly installed preweathered zinc has a smooth, almost glossy feel. This is due to the final passes in the reducing rolls at the mill. These rolls are polished, and they impart a smoothness to the surface. Often, the mills add a light protective coating after the preaging process.

Zinc is very resistant to the surrounding conditions when it has developed the carbonate layer. For all zinc exposed to the atmosphere, the tendency is for the carbonate to form. First, zinc oxide and hydroxide form on the exposed surface as moisture from the air interact. Slowly, over time the hydroxide changes to a carbonate as carbon dioxide in the air and in moisture interacts with the zinc. The initial phosphate layer, applied by some zinc mill suppliers, is a transition coating used to stabilize the surface. It will eventually also absorb carbon dioxide from moisture and transform into zinc carbonate. The zinc carbonate surface provided by other mill sources, gets a jump start on the formation of the zinc carbonate. The coating produced at the mill is very thin.

The patinas are subject to their own special interaction with the environment, but these also are resistant to many substances that would alter the finish appearance. Usually you can see whitish formations on close inspection. This is where the porosity of the patina finishes allows moisture to interact with the zinc surface.

The various exposures encountered by zinc used in art and architecture include relatively consistent and stable interior environments. Or zinc used as a cladding or sculptural form, will experience the extensive variations encountered in the external environment. Coastal, urban environments, rural exposures, or those exposures that fall in between – each pose a unique challenge to the use of zinc. The substrate used to provide support, the time of wetness of both the exposed and unexposed surfaces of the metal, and the cleaning regimen and schedule performed will all impact the performance of zinc in exterior environments.

Performance of zinc depends on:

- Exposure
- Substrate used
- Time of wetness
- Cleaning regimen

260   Chapter 8   Maintaining the Zinc Surface

WHY A MAINTENANCE PROCEDURE

In many instances where zinc is used as an exterior surface, there is only the natural waters from periodic rainstorms that provide the cleaning regimen. In many environments, this works sufficiently well. It is those isolated instances where substances are introduced to the zinc surface that pose, at minimum appearance variations and at worse, deleterious conditions that can lead to failure. Figure 8.2 shows a zinc surface that each winter receives a deluge of deicing salts. Unfortunately, the lack of care or understanding of what these salts can do to a metal surface or any surface, if they are allowed to remain, can ruin an attractive wall or entryway. Simply look at the pitted concrete below the zinc wall and you can see the excessive use of the salt. If the surface was rinsed in the spring this would not occur, but absent this and the salts are doing significant damage to this area. This is the main entryway to the building. These images were taken in June, long after the icing period, and you can still see salt deposits on the wall. Maintenance should include salting but also should include removing the salt in the spring.

The zinc will hold up fairly well to the corrosive attack of salts, but the finish will be damaged beyond repair. It is apparent that there has been little effort to clean this area, and eventually the concrete sidewalk with spall and require resurfacing. The zinc will require replacement. All of which could have been avoided if maintenance plans included a spring rinse down to remove the salt.

DEVELOP A MAINTENANCE STRATEGY

Cleaning and maintaining zinc surfaces are not overly difficult. Similar to other metal surfaces, dislodging staining materials that have not yet keyed into the layers of oxide is the main task. For

FIGURE 8.2   Damage from leaving deicing salts on the surface.

the zinc and region shown in Figure 8.2, a simple, basic process of cleaning the area every spring would have corrected the problem and the ongoing destruction created by the deicing salts. Further in this chapter, a more detailed discussion on deicing salts and their nature will be presented.

The entry ways of buildings in northern climates deicing salts are needed to thaw the ice and avoid a pedestrian from getting harmed. These salts work well on ice in the winter months. In the winter, the cold weather slows chemical reactions down, so the salt is not reacting with the zinc during the time it is working on the ice. However, when the temperatures start to warm up, condensation forms on the metal and any salt on the zinc surface creates a strong electrolyte in the condensed moisture. In fact, the salt is hydrophilic and attracts moisture and holds it. The zinc at the surface reacts to form zinc chloride, a white compound that is partially soluble.

It is a certainty that the building maintenance cleans the interior entryway where people have tracked the salt into the building. They most likely performed maintenance on the interior several times during the winter and as the temperature warms. There should have been a maintenance program to immediately, upon warmer temperatures when the deicing salt is no longer needed, thoroughly rinse the surface of all salt substances.

Thoroughly cleaning the concrete and the zinc wall of all deicing salts each spring would prevent premature damage to the metal surfaces and other surrounding materials such as concrete sidewalks and plants. Fresh water accompanied with pressure is all that is needed for removing the potentially damaging salts.

A maintenance strategy is a valuable tool for a major architectural project as well as a metal art sculpture. The strategy allows for recording and analysis to determine cause and prevention. Every building and art piece occupies a specific place in space, and this leads to unique situations that require planning. One approach to cleaning does not fit all the possible conditions. Therefore, creating a plan for a specific building or piece of art or series of sculpture, is critical to address those unique situations.

A maintenance strategy should include the following aspects:

- Where – Develop a map of the area.
    A simple plan view would suffice note the areas where the deicing salts, human interaction, and other regions that are affected by changing conditions.
- When – Develop a schedule.
    Set out a period where the areas are to be inspected. For deicing conditions, it would be just after winter when the temperatures rise above 10°C (50°F). A schedule should involve

(continued)

(*continued*)

several inspections throughout the year to identify dents, scratches, loose seams and other changing conditions that may present themselves.
- How – Develop a basic equipment inventory.
  Fresh water source and access, power-washing equipment, scrub brush, deionized water, or mild detergent.
- Record – Use logbook or computer file.
  The importance of logging the findings is critical. This gives you a reference to refer to. Changes over time will be noted. Dated images should be included so you can compare from one year to the next. This will also help you define the schedule and note the equipment needed.

A map is a great starting point for a maintenance plan on the metal surfaces. The map shows where human contact is more prevalent, areas that need more periodic review, areas that need more attention after a storm event – areas where debris collect or substances interface with the metal.

The schedule is important. For deicing salts for instance, you must wait until the spring thaw or you can inadvertently create icing conditions with the clean-up. The schedule is a predictive device setting out when a surface is most at danger of harmful substance and addressing it before it occurs. Preventive maintenance is exercised in many other industries such as manufacturing and automotive to undertake necessary cleaning and adjustments before something major occurs. Preventive maintenance, regularly scheduled, is low in cost compared to restorative maintenance and reconstruction.

The methods used to clean must be well researched for the material to be cleaned and the substances to remove. For zinc, avoid using acids or acidic substances, these can harm the oxide coating. Industrial grade detergents, alcohol, organic cleaners such as lacquer thinners, xylene, toluene, and acetone will not affect the oxide surface. However, with zinc, alkaline cleaners can be as harmful as acidic cleaners. Note the graph in Chapter 7, Figure 7.4, cleaners that fall within the range of pH from pH 6 to pH 12 do not harm zinc.

If alkaline cleaners are needed to clean the zinc surface, consider the following solution:

	Dipping or Immersion	Spraying
- Sodium carbonate	10%	20%
- Sodium metasilicate	15%	10%
- Tetra sodium pyrophosphate	20%	65%

This solution will clean soils and grease from the surfaces of zinc and galvanized steel. Rinse thoroughly after cleaning.

Pressure washing the surface can remove loose particles and debris without damaging well established surface oxides. The oxide is very hard and tenacious when formed and will withstand 130 bars (2000 psi) when sprayed from a few meters distance. Use clean water or water and detergent. Do not use abrasives. It is recommended to schedule a thorough rinse of the surface at least once a year to remove debris and other foreign substances that have gotten on the surface.

A periodic wash with clean water and a mild detergent followed by rinsing with clean water performed once or twice a year will go a long way in enabling the surface to resist corrosion from surface pollution. If deionized water is available for the rinse water, it should be used. Deionized water is water that has had the ions removed. Different than distilled water, deionized water will capture the free ions on the surface, particularly salts, and hold them in suspension so they can be rinsed off.

There are additives that can be added to the spray water that aid in removing the chlorides. These are proprietary solutions that essentially turn the water into deionized water to capture free chlorides on the surface.

A light scrubbing to remove road dirt, animal waste, and other detritus that may have become adhered to the surface may be needed. When scrubbing the surface, use a nonabrasive material such as a bristle brush or light nylon brush. You want to avoid scratching the surface. Figure 8.3 shows the scrubbing down of a zinc sculpture to remove deposits that have collected.

Steam cleaning with portable, pressurized cleaners is an excellent method of removing most soils and stubborn bird waste that has calcified on the surface. Detergent can be added, if necessary, to the steam to assist in dislodging dirt and grime. Detergents have surfactants that lift the soils from the surface and allow the pressure of the steam to remove them.

The sculpture in Figure 8.3 is nearly 100 years old. A maintenance program will keep it looking well for another 100 years. Identifying areas that are showing levels of deterioration monitoring them between periods of cleaning will determine if the surface is deteriorating further, the rate of deterioration can provide a comparative analysis to other areas on the surface. If repairs are made, these are kept under observation to determine the efficacy of the repairs.

The importance of a record should not be overlooked. During this light cleaning, an up-close inspection of the surface can be performed to identify areas where corrosion is showing through the coating. Areas of ponding and weakness in coating performance can be noted. Keeping a maintenance log with images and dates is an excellent method to identify areas of concern and monitor any changes they may occur over time. This way strategies can be devised to address the weaknesses before they lead to extensive damage.

Maintaining a Zinc Surface

- Periodic wash with clean water and mild detergent.
- Lightly scrub to remove stubborn waste products.
- Inspect the surface, photograph, and log any issues for review.
- Develop a maintenance strategy.

264   Chapter 8   Maintaining the Zinc Surface

FIGURE 8.3   Maintenance being performed on a large zinc sculpture.
Source: Courtesy of Zahner Metal Conservation

RESTORING THE PREWEATHERED APPEARANCE

The preweathered zinc surface created for rolled-zinc sheet and plate is also generated over time on sculpture made from cast zinc. Once cast, the silvery blue zinc surface is brought out at the foundry by removing mold residues and oxide stains. After this the surface is darkened or allowed to darken naturally. It will take two to five years for the zinc carbonate to form naturally on the surface.

Figure 8.4 shows a cast zinc statue with a darkened surface. The flower had been removed at some point by vandalism. A new flower was recast, then a preweathering treatment by phosphoric

Restoring the Preweathered Appearance

FIGURE 8.4 Zinc statue.

acid was performed to darken the bright zinc of the cast flower. The surface dulled as 60% phosphoric acid was repeatedly wiped over the bright surface until a matching dark tone was achieved. There are other ways of darkening zinc; acetic acid will take the shine off of new galvanized steel surfaces. Treatments involving a copper nitrate and potassium chlorate solution will also darken the metal.

On large, flat surfaces, restoration of the darkened zinc is difficult to achieve without creating a streaky surface. It is recommended to use sponges and lightly treat the surface with dilute phosphoric acid. Repeating this several times will darken the zinc, but it is difficult to avoid streaks on the surface. Follow with dilute acetic acid wipe and then thoroughly rinse the surface.

EFFECTS OF DIFFERENT ENVIRONMENTS

The preweathered surface has excellent corrosion resistance in most environments. The zinc phosphate and zinc carbonate outer layer of corrosion products created by preweathering is very unreactive to most substances in our built environment. The oxide and carbonate that form is an integral extension of the zinc and offers long term and effective protection from what typically would be experienced in the natural environment.

This also goes for galvanized steel with a hot-dipped coating. The zinc outer layer will perform similar to that of the commercial pure zinc. Figure 8.5 shows a galvanized steel support bracket exposed for 25 years in a rural environment. The bracket is attached to a vertical copper wall. For 25 years the bracket and the pipe cladding of galvanized steel were the recipient of moisture running down the copper wall. Corrosion is heaviest at the point of contact. This is galvanic corrosion where the copper is the more noble metal. The zinc has been consumed in its efforts to save the steel, but now the steel is corroding. The corrosion of the steel began to appear after around 10 years' time.

In the right image, the same bracket and fastener type are shown but this is the underside. The zinc is still performing well in this sheltered area where it does not see the condensation. There is deterioration of the zinc, as indicated by the streaking on the copper. The corrosion is very minor after 25 years.

There are many anecdotal examples of similar occurrences; however, coupling zinc with copper is not advised. But the point is that the zinc has some resilience when it comes to the environment humans frequent. The only electrolyte this connection is experiencing is the morning dew that forms on the copper and galvanized steel. The wall faces the east, so it heats up more rapidly and this dries out the surface.

FIGURE 8.5 Galvanized steel bracket attached to a copper wall; 25 years of exposure.

In all cases, the ultimate goal of the zinc surface is to arrive at a sound, inert oxide coating in a reasonable amount of time. As the surface oxidizes, there is no visible corrosion product staining adjacent surfaces. The corrosion products that do come off the surface are minimal and wash away rather than accumulate.

To achieve this sound zinc oxide layer, it is crucial that the surface of the zinc is clean and that it is not the recipient of the oxides and salts coming from other metals. Once clean, and the surface develops the oxide, it is important to maintain a marginal level of cleanliness. Much of this is achieved in good design.

> One caution to consider when cleaning the surfaces of zinc and galvanized steel artifacts or artwork, zinc is amphoteric. This means it is not resistant to acidic substances or alkali substances. The oxides of zinc, as well, are not resistant to the strong acids and strong bases often used as cleaning fluids. Therefore, cleaners used on the zinc surface should be chosen with this in mind. Use neutral pH cleaners, deionized water, and isopropyl alcohol to remove mild contaminants on the surface whenever possible. If acids or bases are used, rinse the surface thoroughly afterward.

Several categories classify a metal surface for cleanliness. Each of these establish the base line as the metal and the form of the metal as it was originally intended.

The categories are:

- Physical cleanliness
- Chemical cleanliness
- Mechanical cleanliness

The following define the basis of these categories in the context of maintaining a surface for art and architecture.

PHYSICAL CLEANLINESS

Physical cleanliness relates to the degree the zinc surface is free of grease, oils, polishing compounds, casting residue, and other soils that are not related to the metal but that can be found on the surface and have an altering effect on the development of the oxide. Fingerprints would fall into this category as would bird waste, adhesives and other foreign substances that detract from the aesthetic and can lead to premature and unplanned changes to the surface. For zinc, if the surface is not kept clean of these substances, it may lead to chemical reactions that can develop long-term irreversible affects.

For most art and architectural uses, physical cleanliness can be achieved on zinc surfaces with milder solvents rather than industrial cleaners. Biodegradable or readily accessible cleaners such as deionized water, isopropyl alcohol, and mild detergents should be the first choice.

There are a number of soils that can find their way to the surface of zinc fabrications. On certain finishes, they are not always that simple to remove. Zinc and galvanized steel fresh from the mill source can often have oils present on the surface. These are protective oils applied to prevent moisture from condensation from damaging the surface during storage. When the oils have been removed, the surfaces will fingerprint quickly as the porous surface captures the oils from your hands.

Mill Oils

Mill oils are light-grade hydrocarbon oils applied at the mill after the final cold-rolling pass. Sometimes there is a light oil applied after the preweathering process to rolled-zinc sheet. These oils are used to combat oxidation during storage and handling. If the steel is destined to weather naturally, these thin oils will evaporate or rinse from the surface at the first rain. Steel has the oil removed as a process step before hot dipping in molten zinc. After the steel is galvanized, the surfaces often receive a light mill oil to protect the fresh zinc surface from oxidation.

On the thin rolled-zinc sheet often an invisible mill oil is applied. This will weather off as the zinc is exposed to the atmosphere.

Die castings will have a coating of mill oil to prevent them from developing the zinc hydroxide in the event condensation or other moisture finds its way the parts during storage and handling. This oil keeps fingerprints from establishing a hold on the surface and resists staining from moisture.

Fingerprints

The first time the new zinc surface is touched by the bare hand, a fingerprint can be seen. The most minor touch can leave a very thin film of human perspiration on the surface because zinc reacts with the oils in hands. This thin film interferes with light reflecting off the surface giving rise to a smudge mark. It remains on the surface until it is displaced with another fluid or it is removed. Fingerprints are composed of organic oils and fats called lipids, amino acids, and water produced by the body. There are often salts intermixed in the oils, but the amount of chloride in a fingerprint is insignificant. The water will evaporate but the oils and fatty substances remain. It is these oils and fats that are so persistent and difficult to remove. They enter the minute ridges and pores on the zinc surface.

If allowed to remain on the fresh zinc surface, these oils and fats will etch the zinc and create an oxide. This usually takes a few days, but the marks left from handling will remain distinctive even as the metal weathers and darkens. Therefore, if the fingerprints have not had the time to etch the metal, they should be removed quickly. Many cleaners just move the oils around or thin them out.

Fingerprints can etch galvanized steel (see Figure 8.6) and can change the natural zinc surface but will not affect the preweathered zinc surface. The art piece shown in Figure 8.6 had been handled

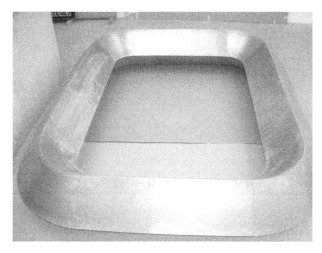

FIGURE 8.6   Fingerprints on thin galvanized steel art form by Donald Judd.

most likely with bare hands. The fingerprints and palm prints are showing along the edge of the galvanized surface. The fingerprints did not show up initially, but they soon etched the surface.

For some patinas, when initially developed or in the process of developing, fingerprints can show as darkened smudges. Fingerprinting is a challenge for many metals used without clear coating or sealing treatments. It is recommended that gloves be used when handling newly galvanized or natural unweathered zinc.

Figure 8.7 shows the before-and-after surface of a natural zinc sheet. Fingerprints are visible on the surface. These have slightly etched the metal and applying mild solvents such as isopropyl alcohol, denatured alcohol, deionized water, or detergents did not remove the fingerprints and smudges.

When the fingerprints become etched into the surface, they can be removed by using a paste containing dilute ammonium hydroxide and oxalic acid. There are commercial metal cleaners available, containing these substances. The pH is around 10, which is not harmful to the zinc if it is removed after application. Household ammonia (dilute ammonium hydroxide), will remove light oxides, but for fingerprints several applications will be needed. It is very important to remove all traces of these cleaning solutions from the surface because if the smallest trace remains, more damaging corrosion can develop.

> Removing handprints and light soils before they have etched the surface
> - Wipe with clean cotton cloth and 99% isopropyl alcohol.
> - Follow with a mild detergent or commercial glass cleaner.
> - Wipe with clean cloth and deionized water.
>
> *(continued)*

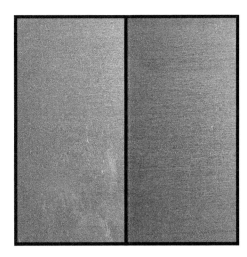

FIGURE 8.7 Fingerprinting of zinc on left. After removal on right.

(*continued*)

Removing handprints and light soils after they have etched the surface

- Use household ammonia.
- Use commercial metal cleaners containing ammonium hydroxide and oxalic acid.

The use of 99% isopropyl alcohol breaks down the oils and smears and gets into the fine grains and displaces the oils. It dries and evaporates quickly, but before it dries out, it has moved the oils closer to the surface. The glass cleaner has a surfactant that lifts and displaces the oils from the surface. Additional wiping down with deionized water will remove any detergent left from the glass cleaner. Ammonium-based glass cleaners will remove the smears and lifted oils, but it is very important to remove any residue and ammonium cleaning fluids.

On preweathered zinc surfaces, fingerprints and smudges can be removed easily with commercial glass cleaners and mild soap and water. Deionized water, followed by a thorough wipe down with a clean cotton cloth, is an excellent way to remove light fingerprinting. On the preweathered surface, the oils from the hands will do little harm. The preweathering oxide will provide a sufficient barrier. It is recommended to wipe down the preweathered surface after installation to remove any potentially harmful substances they may be carried with the fingerprints before they have time to establish a foothold on the zinc surface.

When processing the surface in the plant to develop patinas or other surface finish enhancement, removing oils and grease prior to weathering or patination is required. It can be as simple as a wipe down with acetone, xylene, isopropyl alcohol, methyl alcohols, or mineral spirits.

Rarely, in architectural and artwork preparation, is it necessary to consider hot degreasing or vapor degreasing.

Oils and lubricants come into play if surfaces or castings undergo some level of machining. Lubricants are also used in deep drawing and some stamping operations. Galvanized steels are often coated with a layer of mill oil that wears away over time as the metal is exposed. Until this occurs, the mill oil gives the metal a greasy feel.

Zinc has an advantage over other metals in that it possesses a "self-lubricating" characteristic. This can reduce the need for excessive cutting fluids and thus make final cleanup of the zinc fabrication easier.

Dirt and Grime

Removing dirt and grime from a weathered zinc surface is not difficult. Soap and water perform well and if the grime is very adherent, use steam. These will not have an effect on the weathered zinc surface.

On the patina surfaces, dirt and grime can stick to the coarse surface of the patinated zinc layer. Grime that is oily should be removed using steam or soap and water with a soft bristle brush. Be careful not to remove the patina layer. Bird waste can be especially oily, making it difficult to remove from the surface.

Similar to high-pressure washing, steam will effectively remove baked-on contaminants from the surface of zinc. Steam gets deep into the metal and pulls out dirt and grime from the surface pores. The energy in the steam acts to break down the bonds that hold the grime and dirt to the surface and the force of the steam moves the contamination away from the surface.

Figure 8.8 shows an interior surface of the Roano™ patina zinc that has a visible smear across several panels. This needed to be removed without damaging the patina or replacing the panels.

FIGURE 8.8 Smudges on patina zinc surface used on an interior wall.

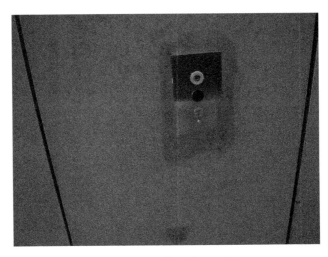

FIGURE 8.9 Stainless steel cleaner affects the surrounding zinc surface.

Solvents such as denatured alcohol, isopropyl alcohol, and deionized water removed some of the mark. The porosity of the patina can make this kind of stain difficult to lift from the surface. In the end, a mild detergent worked the best in removing the stain.

Some smudge marks can be from cleaning fluids used on adjacent materials. Figure 8.9 shows an elevator button made of stainless steel cut into the Roano™ zinc panel. An oily stainless steel cleaner was used liberally and got on the surrounding zinc making a darkened halo around the push button plate.

This did not harm the zinc, but the unsightly appearance needed to be addressed. The use of isopropyl alcohol to lift the oil from the surface and then a light detergent addressed the problem.

When using alcohols, to address cleaning issues on zinc or other semiporous material, one point to consider, is to avoid denatured alcohol. Denatured alcohol is ethanol with an additive to make it nonconsumable. Usually methanol is added to make it poisonous. Some of these additives can leave a film. Isopropyl alcohol usually is diluted with water and does not leave a film.

Adhesives

Adhesives and glues on the surface of zinc can be removed with solvents. Additions of steam can sometimes upset the gummy substance and aid in the removal from the surface. Solvents such as mineral spirits work well in removing adhesives and glues without damaging the patina or weathered oxide layer.

When adhesives remain on the metal for a length of time and bake in the sun, they can change in composition and be very difficult to remove without damaging the surface of zinc. This may necessitate a more powerful solvent, ethyl acetate, or methyl ethyl ketone (MEK) may need to be used with steam. MEK is considered a hazardous substance. Most hydrocarbon-based solvents are

TABLE 8.1   Various Types of Solvents

Hydrogen Bond Donor Acceptor Solvents

Methyl alcohol

Ethyl alcohol

Isopropyl alcohol

Ethylene glycol

Low Hydrogen Bonding Solvents

Naphtha

Mineral spirits

Toluene

Xylene

Hydrogen Bond Acceptor Solvents

Ethyl acetate

Methyl ethyl ketone

Isopropyl acetate

Butyl acetate

flammable and require special handling and disposal as well as special personal safety gear. You may wish to make a poultice to hold the solvent directly on the adhesive for a few hours. This will loosen the bonds and allow the substance to come off by rinsing or steaming.

Table 8.1 shows several solvents and their basic polarity. For example, oxygen in the water molecule is a hydrogen bond acceptor. Different solvents will attract different molecules and break the adhesive bonds that bind them together. It is suggested to start with the top group and work your way down.

Most adhesives used on protective plastics for metals today will degrade and decompose as they are exposed to sunlight and rains. There are occasions where a thin layer of adhesive persists and alters the color tones. Figure 8.10 is a zinc surface where a thin layer of adhesive remained on several panels. Some differences in color remained for a period of time. It eventually weathered from the surface.

Deposits from Sealant Decomposition

Silicone and other sealant material are commonly used to close the joints where one material stops, and the zinc surface begins. It is good practice to avoid the use of exposed sealants on the joints of metal. They collect dirt and stain the adjacent metal surface around the sealant joint.

FIGURE 8.10 Discoloration from adhesives.

Many sealants undergo a catalyzing process as they cure. They contain plasticizers and oily polymers that lead to the development of stains by migrating out of the joint onto the zinc surface. Sealants by their nature are hydrophobic. Water-based cleaners are repelled.

So, the benefits derived from rains rinsing the soils from the surface is lost around sealant joints. A halo surrounding the joint will appear when the surface is wetted. This is an indication of the presence of the thin oil layer.

Pressure washing with hot detergent alone does not remove the stains from sealants, nor will most mild solvents. MEK will dissolve the stain and remove the soil, but this solvent is not recommended due to the safety and environmental hazard posed by this solvent. Ethyl acetate will aid in the removal of the stain if you allow it to sit and work on the surface. Similar to other adhesive stains, creating a poultice to hold the solvent on the sealant stain will work to soften it. Once softened you can rub the silicone from the surface.

Butyl acetate is a similar solvent to ethyl acetate. Butyl acetate will not harm the zinc surface, but its effectiveness is limited. Butyl acetate is used as a solvent for adhesives and some paints.

Grease Deposits from Building Exhaust Systems

Grease that collects on the zinc surface around exhaust systems used to vent kitchens or other food processing can be an unsightly challenge to remove. These hot deposits will darken the metal surface and can enter deep into the surface itself and bake on. This type of stain can be eliminated by proper design and venting.

The difficulty lies in the mass of the oily exhaust. Composed of carbon and fatty substances, these organic soils alight on the cooler metal surfaces and coat the metal with a sticky crust. This greasy deposit is applied warm and can develop into a deep crust containing mold and bacteria as well as the decomposing fats.

Hot steam under pressure with detergent can remove much of this. Deep cleaning with an organic acidic cleaner such as acetic acid or phosphoric acid along with a mild abrasive will aid in dissolving these tough surface deposits, however, this can change surface appearance by removing some of the oxide layer with the grease deposits.

Hydrogen peroxide made into a paste with baking soda (sodium bicarbonate) and allowed to set on the surface for a few minutes can loosen the organic soils. Follow with a steam cleaning to remove the loose particles from the surface.

Graffiti

In the event that a surface is accidentally or purposely painted, the paint can be removed with various techniques. Do not use heat on zinc to remove paint. Avoid using scrapers or any form of abrasive on the surface.

Zinc surfaces will not be affected by the application of any common paint removal solvent. Refer to Figure 8.11. Even strong organic solvents will not affect the oxide or the base metal. It is advisable to use nonalkaline paint strippers. Caustic paint strippers contain sodium hydroxide, need to remain on the surface for a period of time. This can harm the zinc and will alter the appearance of the oxide. The oxides that form on the surface are inorganic, mineralized substances that will not dissolve in solvents used to remove paint but will be affected by the highly alkaline caustic paint removers.

Depending on the circumstances of the area the work must be performed in, health, odor, and safety may play a role in deciding what paint remover to use. The solvent-based paint removers are the most effective but also the more hazardous. They are aromatic and usually contain a combination of the following: xylene, toluene, methanol, and methylene chloride. The odor comes from the

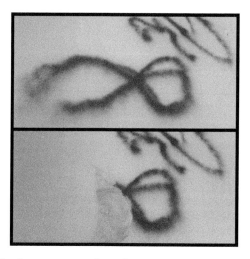

FIGURE 8.11 Graffiti removed using common paint solvents.

fact these are high in volatile organic compounds (VOCs) that evaporate. These will require the user to wear special respirators and gloves. Working around the public will require special venting and protection as well. The benefit with these solvent paint removers is that they work quickly and can address a wide range of paints. They are inexpensive and readily available at most hardware stores.

There are low- and zero-VOC paint removers that use benzyl alcohol, an odorless substance that will dissolve water-based and oil-based paints from zinc without harming the oxide surface. These require more time to break the bonds between the paint and the metal surface. It is still advised to use personal protective gear.

Inks

Metals that are typically destined for industries where surface appearance is not the primary concern, printing designations on the surface to qualify the metal is common.

Galvanized steel is one of those metals that, when produced in coil form, is often destined for less aesthetic end uses. Figure 8.12 shows a galvanized surface with lettering L.F.Q. printed on the surface. This stands for "lock former quality," which was a designation used for a particular grade of steel used in the heating and ventilation industry to make lock formed edges for duct work.

To remove this mark, mineral spirits or other paint removers wiped over the surface is sufficient. They will not have an effect on the galvanized coating. Use a soft cotton wrap and avoid using abrasives.

Concrete and Mortar

These can be difficult to remove without damage to the zinc surface. If the concrete or mortar are still wet, use a pressure hose with clean water to remove it. Do not allow it to set up and harden.

FIGURE 8.12 L.F.Q. Printing on the surface of thin galvanized steel sheet.

TABLE 8.2 Physical Cleanliness

Condition	Cleaning Regimen-Physical Cleanliness
Oils and grease	Detergent, steam, solvent degreasing
Fingerprints	Detergent, glass cleaner, isopropyl alcohol, deionized water
Dirt and grime	Detergent, steam
Etched Fingerprints	Dilute ammonium hydroxide and oxalic acid paste
Bird waste	High-pressure wash, detergent, steam
Adhesives, gums	Solvent
Silicones	Solvent
Building exhaust	Solvent, steam, detergent
Graffiti	Solvent
Ink	Solvent
Concrete or plaster spatter	High-pressure wash. Avoid abrasives. Oxalic acid.

If the concrete or mortar have set up and hardened to the surface, try a solution of oxalic acid. Allow it to set on the concrete then rinse it off with a high pressure rinse. Never use muriatic acid (dilute hydrochloric acid). There will be spots left behind. If they are minor, then leave them. Major spots will require replacement of the zinc.

Tools to Achieve Physical Cleanliness

There are a number of tools at one's disposal to achieve a physically clean surface. Most of these tools are readily available. Physical cleanliness considers the removal of foreign substances that have not damaged the underlying metal or metal surface but instead are adhering or bonding to the surface. Removal requires lifting the substance from the metal and rinsing it away. Some of the tools are listed below.

High-Pressure Water Blasting

Similar to washing your car, high-pressure water blasting is an excellent and inexpensive method of removing substances adhering to the zinc surface. The oxide on zinc surfaces and patinas that have fully developed, will not flake when hit with high-pressure washing. General dirt, bird debris, and fresh concrete can be removed effectively before they can have a detrimental effect on the zinc surface. Removal of salt deposits such as deicing salts can and should be performed each spring.

High-pressure washing should use clean water, with or without detergent. Nonalkaline detergents can be used to lift oils and grease that has not fully hardened. A final rinse with deionized water, if possible, is another added means of removing soils and minute deposits from the surface.

Steam Cleaning

Several commercial steam-cleaning systems operated similar to high-pressure water blast but instead incorporate a small boiler that heats up the clean water and delivers a blast of hot steam. The pressurized steam gets into the metal pores and breaks the bonds holding the foreign substance to the metal.

Steam cleaners operating in the temperature range of 66°–149°C (150°–300°F) work well without overheating the zinc surface.

Deionized Water

Deionized water is an excellent final rinse for metals. Deionized water is water that has had all the dissolved salts removed. Water is passed through a series of tanks containing charged resins, called *ion exchange resins*. Essentially, what occurs are the metal ions dissolved in the water, calcium, iron, magnesium, sodium, copper, are attracted to the anion resin and exchanged for hydrogen ions. The metals are held in the anion tank for later disposal. The anions in the water, chlorides, fluorides, nitrates, and sulfates are removed in the next tank containing cation resins and exchanged for hydroxyl ions. The resulting water is free of metal salts. The process of making deionized water is quicker and more economical than distilled water.

The deionized water attracts other minerals and free elements on the surface of the metal by virtue of the water alone. Since the ions have been removed, deionized water seeks out all ionic particles that can be on the surface of the metal.

CHEMICAL CLEANLINESS

This relates to a surface that is free of oxides, chlorides, nitrides, and carbides that are not related to the base metal or intended to be part of the developed oxide on the zinc surface. In this case, these oxides have formed on the metal by outside additions or unexpected influences. These additions can lead to chemical reactions with the zinc surface and pose long-term performance concerns and appearance issues.

These oxides can also be from the corrosion products of other metals. Rust, for example, coming from a corroding steel element can stain the zinc surface. Other oxides, such as copper sulfates coming from a copper surface draining onto a zinc surface, have a much more damaging effect on the zinc.

One of the more common stains that develop on zinc, in particular natural zinc or newly galvanized zinc, is the white, thick, and powdery stain commonly referred to as *storage stain*. It can also develop on weathered zinc that has been inadvertently exposed to moisture while limiting access to air. Figure 8.13 shows a wall that has been assembled from zinc panels exposed to moisture while temporarily stored.

The challenge to cleaning the zinc surface in order to remove oxides and other deposits is preserving or restoring the original weathered surface. There often is a diffusion of zinc ions from the metal into the oxide, making them more of a part of the surface. Removing these will often take the zinc surface with it, leaving an altered spot or blemish that must weather out. There are several techniques of removing this stain. Often a slight ghosting of the stain will remain.

FIGURE 8.13 Oxide stains on zinc.

Rust Stains

Rust, or iron oxide, generated from a steel assembly near or above a zinc surface will stain the zinc oxide and zinc carbonate surface if allowed to remain. As moisture collects around the rust particles, a reaction will occur at the surface and there will be some combination of zinc–iron and oxygen that will produce a dark stain.

Removing the stain is difficult without affecting the color and tone of the zinc. To dissolve some of the iron on the surface, a phosphoric acid treatment or a dilute oxalic acid treatment can be used to remove the stain with the help of a light abrasive. The phosphoric acid will darken the surface; however, the abrasive can remove some of the oxide and zinc carbonate that underlies the stain. This will lighten the surface.

If severe, replacement of the zinc with a matching color may be necessary because of the way the repair has altered the appearance. Figure 8.14 shows a small zinc casting used in a fountain. The fountain piping is corroding, and this constant spray of iron oxide stained the sculpture and created small pits over the surface.

If the stain is minor, consider working delicately with the oxalic acid to remove the reddish brown color and rinse the surface. Address the area where the rust is being generated from and leave the surface to weather out. Once you stop the contributing factor of the rust and remove most of the stain, it will not return or damage the zinc appearance.

Hard-Water Deposits

Hard-water stains fall between physical and chemical cleanliness. If they are simply on the surface and require removal, they would be considered as physical additions to the surface. If, on the other hand, they have engaged into the underlying zinc surface, they become a chemical issue.

FIGURE 8.14 Small zinc statue with severe rust staining.
Source: Image courtesy of Zahner Metal Conservation.

FIGURE 8.15 Zinc sculpture of Neptune. Before-and-after stain removal.

Zinc statues or surfaces that have the constant flow of water can develop stains from the calcium and other substances dissolved in the water. Iron oxide, as seen in Figure 8.15, can damage the surface appearance of zinc and actually etch into the metal. Hard water contains calcium and magnesium, as well as carbonates and sulfates. Additionally, in fountains, there may be added chlorides or bromides to act as an algaecide and keep the water looking clear. Figure 8.15 shows a large zinc sculpture with years of hard-water staining.

Pressure washing, along with some light scrubbing and a quick wipe down with dilute phosphoric acid followed by a thorough rinse, were sufficient to remove much of the stain.

In this case, the water deposits had not reacted with the zinc carbonate surface and came off without damaging the underlying zinc patina. Some whitish deposits, potentially zinc chloride from the algaecide used, have combined with the zinc on the surface.

Storage Staining

The zinc surface will develop a thick, white corrosion product of zinc oxide and zinc hydroxide when stored in such a way that moisture is trapped against the surface and air is prevented from reaching the surface. Known as white storage stain, this can happen in a matter of days. The moisture can be from condensation developing on the surface and becoming trapped in crevices or between two adjacent surfaces or it can happen if the surface is wetted and moisture is pulled between two adjacent surfaces by means of capillary action. Moisture can also become trapped in packaging material and held against the surface.

When moisture is held against the surface and atmospheric carbon dioxide is prevented from interacting, the zinc will pull oxygen out of water and form zinc oxide. Soon after this occurs zinc hydroxide will thicken and swell into a powdery white mass. Figure 8.16 shows a galvanized steel surface after months of storage in a wet environment.

Water streaks and water stains can be unsightly, but they do not effect mechanical or corrosion resistance of the zinc. This whitish stain can occur on the new zinc surfaces such as freshly cast surfaces, wrought sheet surfaces and the newly galvanized surface. The stain does not normally develop on aged zinc surfaces that have developed the zinc carbonate; however, improper storage where the surface is not allowed to drain or where the surface has limited access to air, the stain can appear.

Often the fabricated zinc panel is coated with a plastic film. Refer to Figure 8.17. For most applications the film is required to protect the easily scratched surface from marring during fabrication

FIGURE 8.16 Zinc oxide and hydroxide stain on a galvanized steel surface.

FIGURE 8.17 Image of stains on preweathered.

processes as harder steel tooling is brought against it. The complication occurs during the time the fabrication is complete, and the product is established on the building surface. The product, with the plastic film is stored where it can get wet. Water can wick between the plastic and the zinc surface, all the while the plastic also inhibits air flowing to the surface. If left this way for a few days, the white oxide can form on the contrasting dark preweathered surface creating a visual nightmare.

Fresh patinated surfaces can develop a similar oxide stain if water is permitted to set on the surface. This stain is not the white water storage stain, but a stain due to a concentration of moisture. Dew and condensation collect on a flat surface and run down the face creating a streak of white zinc hydroxide. Figure 8.18 shows a surface with streaks running down the face.

FIGURE 8.18 Hunter patina with an accumulation of zinc hydroxide on the lower face.

This happened early on in the exposure and has slowly grown thicker as moisture runs over the face and the zinc hydroxide grows. The image on the left is after 5 years of exposure and on the right is after 10 years.

Water and storage stains are difficult to remove. There are several techniques that have had measures of success in removing the stain. Laser ablation is one method but may not be suitable for finished zinc surfaces. If you are matching adjoining nonstained zinc surfaces, this can be a challenge. You will need to restore the finish and this effort may be too expensive with unknown results. Once removed the zinc below the stain is lighter, and if galvanized, the spangle is less defined, and the gloss is matte due to microscopic etching of the surface.

Chemical applications have measures of success. One method involves application of 10% acetic acid applied several times. This removes some of the stain but darkens the surface. A bristle brush can be used to assist in the removal.

Another chemical treatment involves:

- 19 milliliters of ammonium hydroxide (household ammonia is sufficient)
- 6 grams of ammonium chloride (sal ammoniac)
- 6 grams of ammonium carbonate (Baker's Ammonia)
- 71 milliliters of water

Apply using a moistened cloth and work in with a soft bristle brush.

A third treatment uses glycolic acid 15% concentration, along with sulfamic acid 5%. Apply a small amount of the solution to the stain and work into the stain. Use a bristle brush to gently break up the thick oxide.

When applying any of these solutions wear personal protection for hands and eyes. Ammonia fumes can by hazardous and it is advised to work in a well ventilated space. Rinse the surface thoroughly to remove all traces and dry the surface to remove most traces of moisture.

Deicing Salts

Deicing salts used to thaw ice on roadways and sidewalks in northern regions is not a friend to most metals used in art and architecture. With zinc, the damage from salt is more cosmetic than metallurgic. Salt deposits are hydrophilic – they hold moisture. The moisture can develop on the metal surface each morning as dew forms on the cooler metal. Figure 8.19 is a diagram that depicts what occurs with a metal surface as the ambient temperature changes throughout the day.

Each time the surface becomes moist, the salt partially goes into solution. This is a strong electrolyte and can set up localized corrosion cells. Such cells can generate small pits in the surface.

Figure 8.20 shows a surface of zinc that has been exposed to salting around the entryway of a building. The salt has been on the surface for months.

These stains are a chlorinated zinc hydroxide. The corrosion product will not damage the zinc unless it is allowed to remain wetted. Vertical surfaces dry out, but the stain remains.

The salts will create discoloration on the surface and combine with the carbonate or hydroxide on the surface. This stain is difficult to remove without damaging the weathered surface of

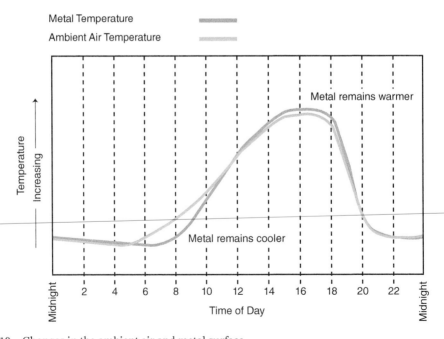

FIGURE 8.19 Changes in the ambient air and metal surface.

FIGURE 8.20   Deicing salts creating minor discolored oxide stain.

zinc carbonate. Further, the surface can be damaged by deicing salts if they are allowed to remain for prolonged periods where they continually are wetted. Moisture with the deicing salts combine to make an excellent electrolyte. Airborne pollutants will combine to create a condition that can develop into corrosion cells.

The damage will be superficial but cosmetic. There will be a stain on the surface as the chloride interfaces with the zinc. The crust can be removed but a whitish stain will remain.

There are several chloride salts that are used for deicing. Sodium chloride is the main salt deposit in seaside environments. Whereas in deicing, sodium chloride is usually mixed with other chlorides. Sodium chloride for deicing is usually mined rather than taken from the sea. The mined salt, known as the mineral halide or the common name rock salt, contains numerous other trace compounds. It is the most economical yet the least efficient for deicing purposes.

De-icing salts work by lowering the freezing point of water. Sodium chloride is an endothermic salt which means it must pull energy from the surroundings. Its effective working temperature is 7°C (20°F), so below this temperature it does little good in melting ice.

Another common salt used is calcium chloride. Calcium chloride lowers the working temperature further by creating an exothermic reaction. It releases heat as it goes into solution. Calcium chloride is also hygroscopic as it can access moisture from the air. Often the deicing salt is provided in a combination of 60–80% sodium chloride and 20–40% calcium chloride.

Magnesium chloride is being used more frequently in many of the deicing mixes. This salt also reduces the effective working temperature and has less of an environmental impact because it is provided as a hydrate. This compound is composed of more than 50% water. Magnesium chloride is exothermic. As it goes into solution it will generate heat. With zinc, the magnesium will have an offsetting affect as the magnesium ion will help resist corrosion of the zinc surface.

TABLE 8.3 Working Temperatures for Deicing Salts

Salt	Temperature
Sodium chloride	7°C (20°F)
Magnesium chloride	−18°C (0°F)
Calcium chloride	−32°C (−25°F)

Various additives have been used, many of which are organic in nature. These additives make the deicing salts sticky when they are in a slurry form, so they adhere to the road and walk surfaces. They also make the deicing salts stick to metal surfaces. Beet juice is a common organic additive.

All of these salts corrode metals when the right conditions are met. Salts, both deicing and coastal, are hydrophilic. They absorb moisture from the air and hold onto it longer. Table 8.3 shows several of the most common salts used and temperature where these salts tend to work to depress the freezing point.

Coastal Exposures

Similar to what occurs with deicing salts, zinc in proximity to the coast will develop a white zinc chloride deposit. The more temperate the climate, the more condensation will occur, and the condensation will capture some of the zinc salts and deposit them at drip edges or offsets. Figure 8.21 shows the eave where salt deposits are collecting around the seam. The bottom image is a close up of the horizontal seam in the eave. The project is in the Bahamas, so salt from the seaside environment and humidity are present.

FIGURE 8.21 Coastal exposure. Zinc roof showing white dusting of zinc chloride.

TABLE 8.4 Composition of Seawater

Approximate % by Dry Weight	Element
55%	Chlorine (Cl)
31%	Sodium (Na)
8%	Sulfur (S)
4%	Magnesium (Mg)
1%	Calcium (Ca)
1%	Potassium (K)

Table 8.4 shows the typical composition of sea water. The components are similar to those in deicing salts and the effect on the surface is the same. There is a buildup of white zinc chloride deposits that are partially soluble in water. Some of this will rinse away while some of the salt collects and adheres to the surface of the zinc. The color of the zinc overall is lighter in color than a zinc surface in a northern climate where the zinc carbonate is the prevalent oxide on the surface.

Construction Dust and Debris

Construction sites can be rough on zinc surfaces as other trades process their work and foreign substances come in contact with the surface of zinc. These can amount to any number of things from concrete dust, steel particles, fireproof overspray, sealants, and so on. All articles of construction that can find their way to the finish surface of zinc. Most of these substances are benign and can be removed without damage if addressed quickly. A few substances (e.g., wet concrete, fireproofing overspray, bitumen, and sealants) can adhere to the surface and may need some level of physical removal. The challenge is to not damage the surface oxide layer.

Figure 8.22 shows a patinated surface with white stains from concrete wash and dust. The patinated zinc surface can be the most delicate to clean because it is coarse; however, similar issues can develop from the preweathered surfaces when substances are allowed to remain on the surface.

In all cases it is good practice to address the cleaning immediately to avoid chemical interaction of the foreign substance with the outer layer of zinc. The concrete dust can be rinsed from the surface. It may take several attempts but pressure rinsing, at low bar pressure should remove the dust.

On wet concrete, fireproofing overspray, and wet plaster, use a high-pressure spray without abrasive to remove the substances. You can add a fiber bristle brush – but only carefully, to avoid scratching and marring the surface. If allowed to remain on the surface for an appreciable amount of time, replacement of the zinc elements may be the only alternative.

Sealants and bitumen can leave an oil residue behind. To remove sealants and gums, use a knife or razor to remove excess lumps down as far as possible without damaging the zinc surface. Use mineral spirits to dissolve as much as possible and step up to methyl ethyl ketone to remove the last remnants. Rinse the surface thoroughly on completion.

288　Chapter 8　Maintaining the Zinc Surface

FIGURE 8.22　Construction debris wash on patinated zinc surface.

Other Stains

Zinc is used in everyday life for such things as die-cast carburetors on small vehicles, hardware, toys, light fixtures, and various automotive accessories. For these products, the metal is exposed to conditions not considered in art and architecture. In the context of art and architecture it is mainly the exposure to other metals, atmospheric pollutants, and cleaning procedures that can lead to staining and surface degradation.

Zinc is sometimes used in the interior spaces of buildings, such as countertops, tables, and bar tops. Today, the metal is seeing an increase as a table or bar top, reminiscent to the old European pubs. Wine, in particular, can create a stain on the surface of counters or tables made of the metal. Wine is acidic, pH in the range of 2.9 to 3.9, and will interact with the surface and stain the zinc. The stain can be removed by rubbing the surface down with the same formula to remove the water stain. In English pubs, it is said they wipe the surface down with bitters. In any event, this will give the surface an old hand-rubbed metal feel with some lighter areas and some with darker tones. If abrasives are used, then consider only the light household powder cleaners. These will scratch the surface and lighten it. You can redarken by wiping with vinegar. Table 8.5 is a list of various chemical conditions that a zinc surface may encounter and some basic cleaning recommendations.

Tools to Achieve Chemical Cleanliness

Consider storing the zinc articles with adequate space to allow air to reach the surface. Similar to rust on steel, the zinc surface gains a level of sensitivity to further exposure. The following discussion addressees additional ways to achieve chemical cleanliness.

TABLE 8.5 Chemical Cleanliness of Zinc Surfaces

Condition	Cleaning Regimen-Chemical Cleanliness
Rust stains	Mild acids- phosphoric, citric or oxalic acid solutions. Rinse.
Hard-water stains	Pressure wash, mild acid treatment, light scrubbing
White storage stain	Ammonia, ammonium chloride, ammonium carbonate solution
Deicing stains	Clean water rinses every spring. Design and maintenance practices that keep the salts away from the surface. Ammonia, ammonium chloride, ammonium carbonate solution, pressure wash.
Constant wetting	Design to eliminate. Bituminous paint. Venting.
Misc. stains	Ammonia, ammonium chloride, ammonium carbonate solution.
Concrete and plaster – wet	Pressure wash and light scrub. Avoid marring the surface.
Concrete and plaster – dry	Pressure wash, scrub. Oxalic acid.
Sealants, gums, bitumen	Mineral spirits, methyl ethyl ketone

Laser Ablation

Laser ablation removes material from the surface by thermal shock and vaporization. The advent of the fiber laser development has made this technique a viable form of cleaning a surface that has heavy oxides and stains.

Laser ablation is a cleaning method that can be used to remove heavy zinc hydroxide, iron oxide stains and other substances that have developed or attached to the zinc surface. The surface finish, however, will be etched by the laser and will require a restoration of the original texture and surface color.

Essentially, the energy of the laser light is absorbed by the surface contaminants, and this absorption of high energy vaporizes the substances without damaging or effecting the mechanical properties of the zinc. The laser beam is pulsed to the surface and a tremendous amount of energy is released in a short period of time. This energy is absorbed by the particles on the surface of the metal, even the oils of fingerprints. The particles become excited and vibrate as the bonds between the substances and the zinc are broken. A flowing gas sweeps the particles from the surface and a suction pulls the particles into a collection filter. In this way, the surface contamination and oxides are collected and disposed of without release to the atmosphere or redeposited on the ground. This also protects the lens of the laser device.

In laser ablation, the laser itself is kept remote and the beam of light is transferred via a fiber cable. The cable length can be more than several hundred feet if necessary, without affecting the power at the beam to metal contact. The portability afforded by the fiber laser allows for cleaning remote and difficult regions in situ.

The most common fiber laser used for laser cleaning and ablation is a Nd:YAG. The Nd:YAG is more efficient than a CO_2 laser and they are easy to handle and operate. The Nd:YAG is a solid-state laser based on the rare earth doped crystals of yttrium aluminum garnet (YAG). These are manmade crystals specifically designed for lasers. The Nd:YAG uses a doped fiber to deliver the high-power energy to the laser head. For ablation cleaning the Nd:YAG is normally a pulse laser with an emission wavelength of 1064 nm. There are several alternatives to the Nd:YAG that utilize other rare-earth elements and the development of these lasers is advancing rapidly. But the majority of ablation operations are performed with the Nd:YAG fiber system.

For laser ablation to work, the material's wavelength absorption must be compatible with the laser wavelength. The energy needed to break the bonds of the oxides to the surface and the detritus adhesion is below the threshold of the bonds that hold the metal together. When the bonds that hold the contamination are broken the contamination vibrates off of the surface from the excited, high energy electrons.

FIGURE 8.23   Laser ablation on zinc surface with white hydroxide.

Chemical Cleanliness

Unlike other methods of cleaning, no waste is released to the environment, the process requires no solvents, applies no water and leaves the surface free and clean of most all contamination. Laser ablation can be used on wet or dry surfaces.

Special training and care need to be exercised by laser ablation cleaning of zinc surfaces. The skill and expense are higher for this cleaning technique than other chemical techniques. However, the residual waste and environmental impact are significantly lower with this cleaning technique.

When using laser ablation, the surface of the zinc is altered as you can see in Figure 8.23. The surface is etched due to melting at a micro level. The pulsing laser leaves behind a fine etched finish that is different than other mechanical finishes. The process of laser ablation can be performed robotically or manually.

Prevent Storage Stain

To prevent storage stain, keep the surface of the zinc dry as possible. On finish products, store in protective crates until ready. Use pine for the crate material or line the crate with polyethylene to keep out gassing of the wood in the event the wood become moist. If storage is to be any length of time, tilt the crates to drain any moisture or condensation. Vent the crates, allowing air to flow over the surface of the zinc keeping it as dry as practical. Figure 8.24 touches on a few basic suggestions in protecting zinc panels when packaging and shipping may expose the zinc to moisture.

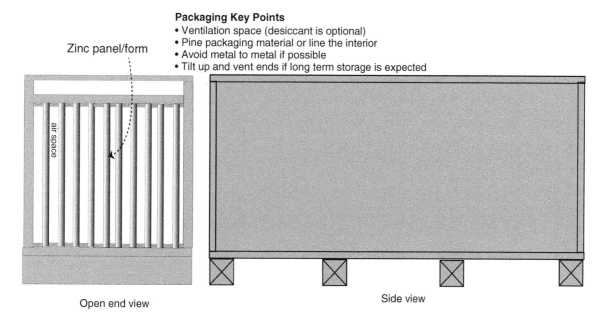

FIGURE 8.24 Example of packaging for zinc products.

MECHANICAL CLEANLINESS

By mechanical cleanliness, a zinc surface must be free of regions that are torn, scratched, dented, and gouged. This would also include surface marring from manufacturing processes and packaging issues that alter the surface. Mechanical cleanliness is a restorative process where the surface regains most if not all of its intended geometry.

Scratches and Mars

Zinc is a soft material and will scratch if not handled properly. The scratch looks dark against the preweathered surface, but at certain angles the bright line of the zinc is apparent. It will age over time and the color will blend but the scuff will persist. Even though rolled zinc does not necessarily have a grain induced from satin polishing, it does have an apparent grain and scratches and mars that cross the grain are more apparent. One method that will work to reduce the visibility of a mar is shown in Figure 8.25.

Figure 8.25 shows a mar on the surface of a preweathered zinc sheet with a visible grain. The mar or scratch is light, but it creates a contrasting color. Going from top left to top right, the repair of the mar involves first putting a light scratch grain in the direction of the sheet grain removes the scratch but lightens the appearance by removing the preweathering.

You can darken the new surface and remove some of the contrast by wiping zinc phosphate mixture such as a Parkerizing solution on the mar or scratch. Phosphoric acid works as well but can darken the surface rapidly. It is recommended to test the surface to ensure you do not make a bad situation into a worse situation. The zinc phosphate creates a matte surface and you apply it sparingly, follow with a wet cloth to wipe off excess.

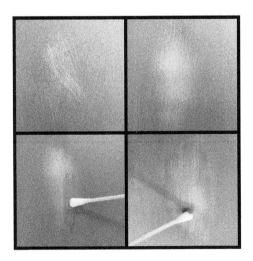

FIGURE 8.25   Mar on zinc surface before and after treatment.

FIGURE 8.26 Scratch across panels during installation.

Repairing scratched or marred surfaces on the patinated zinc sheet is less challenging than repairs on preweathered or natural zinc surfaces. Figure 8.26 shows a scratch into a prepatinated surface during installation. Use a cotton swab with some patination solution supplied by the company that patinated the zinc. Very carefully apply the solution into and over the scratch. Do this very carefully and wipe excess solution off with isopropyl alcohol. Repeat this several times until the scratch blends in. If the scratch is deep, you may want to carefully sand the area to take down the rough edges on the scratch. Then apply the solution with a cotton swab. Remove all excess chemical and follow with dabbing with a sponge saturated with clean water.

Sometimes just wetting the scratch will work to blend it in. This patina is thick and will cover most minor scratches.

Dents

The lower yield strength of zinc can make impact damage a concern when using thin sheet metal forms in high traffic areas. Removing the dent is difficult. If you can access the reverse side the dent can be hammered out. The metal is soft and easy to work if the dent is not severe.

Hail Damage and Small Dent Repair

Hail can dent thin zinc roof and wall surfaces. The lower reflectivity of weathered zinc helps conceal the damage, making it less visible.

Small, rounded dents can be removed. For the dent to have occurred, the zinc would have gone through localized plastic deformation. There may be some thinning as the dent stretches the metal slightly, but this is hardly noticeable. If the dent is acute and has a visible crease, the crease will not be removable. If the dents are marks in the surface such as hammer marks; these as well are not repairable. Smooth concave and convex dents are repairable.

To remove a dent, first lightly abrade the reverse side of the panel or sheet. This will pinpoint exactly where the dent is, and it will highlight the area. Block the face side with a clean, smooth, wooden block, end grain against the metal. On the reverse side, the dent can be carefully hammered out with another wood block against the metal. Work it gently and the dented metal will return close to the original position.

If you cannot access the reverse side, replacement may be the only option. Unlike copper or stainless steel, zinc cannot be heated and then rapidly cooled. You cannot stud weld and pull the dent out. One can try a dent puller with suction to pull rounded dents from the surface.

Heat Distortion

Distortions created by soldering and welding processes are not usually an issue with zinc. Zinc has a high coefficient of expansion and is difficult to join by heat processes, but there is such a relaxation that occurs when the metal is heated, warpage is contained. Heat joining methods require special care to avoid melt out of the metal being joined. Distortions may occur localized around the weld where the metal has drawn in and created a concave distortion. Good welding practices and welding skill will eliminate this.

TIG welding processes, also called GTAW for gas tungsten arc welding, will give good results when welding zinc. The use of an inert gas to keep the weld clean and oxide free is important. Alternating current is the preferred setting. Porosity in the weld is a greater concern than distortion of the metal around the weld.

Distortions from Cold-Forming Operations

Zinc steel sheet and plate are provided levelled and flat. The low strength of these alloys requires less power to shear or form and work hardening does not occur. In the past, stamping of heated zinc sheets to create intricate sculpture was performed. Today, zinc is still stamped from warm sheet into intricate forms to match ornamentation or create ornamentation. The architectural style is less common today.

Creep Dimensional Changes

Creep is a condition where the metal elongates and undergoes geometric changes. Large castings of zinc are subject to creep over time. The weight of the metal under moderate temperatures can elongate and crack the thick zinc casting. Figure 8.27 shows cracks in a large zinc sculpture that have developed over time. The sculpture is nearly a century old. Remediation has been attempted to seal the crack and keep moisture out of the hollow form.

In zinc sculpture made from high-purity zinc, creep is a condition that is difficult to prevent. The cracks form as the material expands and contacts and as the weight of zinc over time stress the areas of weakness. A crack will start and begin to propagate across the piece.

Creep is less a concern for sheet and plate made from alloys of zinc containing copper and titanium. Cast zinc alloys containing aluminum are typically smaller in size and creep stresses do not build up. It is the large cast zinc pieces that can experience this destructive condition. The repairs are difficult to undertake. Figure 8.28 shows a cast zinc sculpture with a split that occurred on the leg of the dog. The crack was filled with mortar on the assumption this would be more the color of the zinc. Other repairs involve soldering the joint to seal the joint. When this can be done, the color and surface match is improved. Sealants are also used.

All of these solutions are designed to keep moisture out. This will help deter destruction from freezing and splitting the sculpture, but ultimately, the creep movement will continue and, over time, the cracks will grow and expand.

FIGURE 8.27   Cracks in zinc sculpture from creep.

FIGURE 8.28   Cast zinc sculpture with crack in dog's leg, filled with mortar.

TABLE 8.6   Mechanical Cleanliness

Condition	Cleaning Regimen-Mechanical Cleanliness
Scratches and mars	Light abrasion follows with phosphoric treatment or with patina solution.
Dents	Block and hammer to reverse damage.
Heat distortion	Use TIG welding and AC current.
Distortion from fabrication	Design and skill. Block and hammer. Replace.
Creep	Keep moisture from entering the crack. Solder and weld if possible. Creep will continue. Control damage from other means such as freezing.

GALVANIZED STEEL SURFACES

Galvanized steel surfaces are often left unpainted to exhibit the character of the zinc crystal, the spangle that develops on thin steel sheet. On exterior surfaces, these are generally allowed to age and weather naturally. The surface will lighten as the zinc oxide changes to zinc hydroxide, then darken out slightly as carbon dioxide is absorbed and zinc carbonate forms. Gloss and reflectivity give way to a matte, low-reflective surface. Once this happens, the surface is resistant to corrosion. Fingerprinting is not an issue; the surface loses its porosity, and with it, the affinity for the oils from your hands. Deicing salts have little consequence unless they create an electrolyte bridge to the underlying steel at cut edges or weak areas. This can happen if deicing salts and moisture are allowed to set on the surface of the galvanized steel.

The thickness of the galvanized coating is the critical parameter for long-term performance of the coating. There are various thicknesses of zinc coatings used on hot-dipped surfaces. Typical hot-dipped coatings applied by batch galvanizing are from 45 μm to as much as 300 μm. The thickness is a function of base steel thickness and time in the molten zinc. It is recommended for exterior exposures to use a thickness no less than 55 μm average, with a minimum spot check of 45 μm. This would be for steel thicknesses 1.5–3 mm. Above this steel thickness, the average should increase. Figure 8.29 shows various thicknesses of galvanized steel and the expected duration of time before the surface shows a significant level of corrosion. Different exposures are indicated. This data is taken from information presented by the American Galvanizers Association.

Zinc coatings eventually will wear as they are consumed by the oxygen-reduction process involved with protecting the base metal. The underlying steel will begin to corrode as the surrounding zinc continues to oxidize. The initial appearance is spotty, but eventually it broadens out. This is what is depicted in the graph shown in Figure 8.29. In different environments you can expect the zinc to hold up according to a correlation to the zinc thickness.

The galvanized surface will perform in the various exposures similar to the natural finish of zinc. The main difference being the characteristic spangle of the zinc crystal is not available in cast and rolled zinc. The processes of creating the surface are very different. Galvanized surfaces develop as steel is immersed in a molten bath. Some of the zinc bonds and the steel surface diffuses into the zinc at the interface. The longer the steel is immersed in the molten zinc, the thicker the zinc coating becomes. When the steel is brought out of the molten zinc bath, the zinc on the surface rapidly cools

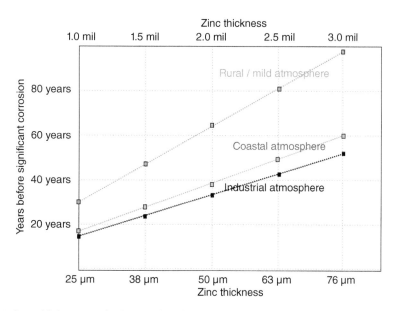

FIGURE 8.29 Various thicknesses of galvanized and the expected lifespan.

FIGURE 8.30 Hot-dipped galvanized steel sculpture by R+K Design.
Source: R+K Design

and solidifies. The spangle or crystals of zinc are generated from single points on the surface. These points are molecules of lead or antimony on the steel surface that solidify first. Chapter 3 discusses this in more depth.

When immersed in high-purity zinc, the coating can form on steel and the spangles are less visible. Figure 8.30 shows a carved and welded steel surface immediately after it has been raised from the molten zinc bath.

When the zinc coating is damaged or scratched during the initial handling and installation, the approach often taken is to apply a coating of zinc-rich paint. Initially the appearance is satisfactory but after a short period of exposure, the paint darkens in comparison to the hot-dipped galvanized surface. It is better to remove all excess paint and leave only a small amount within the scratch.

If not, the touched-up galvanized surface will look splotchy from the differential weathering that occurs.

Once the zinc gives way and the base steel begins to corrode, the only solutions are to recycle the form or paint the surface. You will need to remove the corrosion first. If the corrosion is severe, there will not be sufficient steel to coat with paint.

Whether the surface consists of preweathered zinc carbonate, pure zinc, or patinated zinc, maintaining a minimal level of cleanliness will give decades of beautiful life to the metal.

Artifacts and surfaces constructed of zinc are natural and beautiful. If maintained, zinc can and will perform for generations to enjoy.

APPENDIX A

Brand Names

This is a list of some of the brand names for rolled- and cast-zinc products over the years. Several of these are still in use, while others have been discontinued or the company no longer exists. The effort to brand the zinc products has kept it out of consideration as a commodity, unlike copper, aluminum, or steel.

Tatanalloy
Microzinc Alloy 190 (Ball Metal and Chemical Co)
Zilloy 15 (New Jersey Zinc Company)
Zilloy 20 (New Jersey Zinc Company)
NJZ Alloy No 25 (New Jersey Zinc Company)
Microzinc Alloy 700 (Ball Metal and Chemical Co)
S.T.Z. (Billiton Metals and Ores)
Rheinzink™
Classic (Rheinzink™)
prePATINA (Rheinzink™)
Granum (Rheinzink™)
Jarden
ARTAZN™
Glacier Gray (ARTAZN™)
Onyx (ARTAZN™)
Zintane 2 (Societe des Mines et Fonderies de Zinc)
Zintane 20 (Societe des Mines et Fonderies de Zinc)
RCA – Titane (Compagnie Royale Asutienes des Mines)

SAMIM Venezia (Stablimento die Porto, Marghera)
Titanzinc (Cinkarna)
Natural (ElZinc™)
Slate (ElZinc™)
Graphite (ElZinc™)
Quartz (VM Zinc)
Anthra (VM Zinc)
Pigmento (VM Zinc)
New Nova (Nedzink)
Noir (Nedzinc)
Alloy 190 (Jarden)
Alloy 500 (Jarden)
Alloy 505 (Jarden)
Alloy 710 (Jarden)
Hunter™Zinc (Zahner)
Roano™Zinc (Zahner)
Baroque™Zinc (Zahner)
No. 3 Casting Alloy
Zamak 3
Mazak 3
No. 5 Die Casting Alloy
Zamak 5
Mazak 5
No. 7 Die Cast Alloy
Zamak 7
Mazak 7
ZA 8
ILZRO 12 (International Lead Zinc Research Org.)
ZA 12
Super Z300
Formetal 22 Alloy
Korloy 2684

APPENDIX B

Select Specifications for Zinc

ASTM

B 6	Zinc
B 69	Standard Specification for Rolled Zinc
B 86	Zinc Alloy Die Casting
B 852	Zinc Alloys for Continuous Hot-Dip Galvanizing of Sheet Steel
B 860	Zinc Master Alloys for Use in Hot-Dip Galvanizing
B 960-08	Prime Western Grade-Recycled (PWG-R) Zinc
B 240	Zinc Alloys in Ingot Form for Die Casting
B 327	Master Alloys Used in Making Zinc Die-Casting Alloys
B 792	Zinc Alloys in Ingot Form for Slush Casting
B 833	Zinc Wire for Thermal Spraying (Metalizing)
B 860	Zinc Master Alloys used for Hot-dipped Galvanizing
B 275 -80	Codification of Certain Non-Ferrous Metals and Alloys, Cast and Wrought

Coatings

ASTM A879	Standard Specification for Steel Sheet. Zinc-Coated by the Electrolytic Process
ASTM A653	Standard Specification for Steel Sheet. Zinc-Coated (Galvanized) or Zinc–Iron Alloy-Coated (Galvannealed) by the Hot-Dip Process
ASTM A123	Standard Specification for Zinc (Hot-Dip Galvanized) Coatings on Iron and Steel Products
ASTM A153	Standard Specification for Zinc Coating (Hot-Dip Galvanized) on Iron and Steel Hardware

European

EN 501	Roofing products from metal sheet. Specification for fully supported roofing products of zinc sheet.
EN 612	Eaves Gutters with Bead Stiffened Fronts and Rainwater Pipes with Seamed Joints Made of Sheet Metal.
EN 988	Zinc and Zinc Alloys. Specifications for Zinc and Zinc Alloys Rolled Flat Products for Building.
EN 1179	Zinc and Zinc Alloys. Primary Zinc. Zinc Composition for Zinc Traded Internationally.)
EN 1774	Zinc and Zinc Alloys. Alloys for Foundry Purposes.
EN 12844	Zinc and Zinc Alloys. Alloy Casting Specifications.

References

ASM International (1990). *ASM Handbook, Volume 2: Properties and Selection: Nonferrous Alloys and Special-Purpose Materials*. ASM International.

ASM International (2005). *ASM Handbook, Volume 13B: Corrosion: Materials: Corrosion of Zinc and Zinc Alloys*. ASM International.

Bringas, J. E., and Wayman, M. L. (1998). *The Metals Red Book: Nonferrous Metals 2nd Edition*. Edmonton: Casti Publishing Inc.

Chawla, S. L., Gupta, R. K. (1993). *Material Selection for Corrosion Control*. Ohio: ASTM International.

Coburn, S.K. (1978). *Atmospheric Factors Affecting the Corrosion of Engineering Metals*. Baltimore: ASTM.

Costa, V. (2019). *Modern Metals In Cultural Heritage: Understanding and Characterization*. Los Angeles: Getty Conservation Institute.

Costa, V. (2019). *Modern Metals in Cultural Heritage*. Los Angeles: Getty Conservation Institute.

Cowie, G. (1984). *Designing in Zinc*. New York: International Lead Zinc Research Organization, Inc.

Dean, S. W and Rhea, E. C. (1982). *Atmospheric Corrosion of Metals*. ASTM International, Baltimore.

Draper, M., Draper W. (1946). *Old Grubstake Days in Joplin: The Story of the Pioneers Who Discovered the Largest and Richest Lead and Zinc Mining Field in the World*. Haldeman-Julius Publications, Girard, Kansas.

Grissom, Carol A. (2009). *Zinc Sculpture In America 1850–1950*. University of Delaware Press. Newark.

ICOMOS France (2012). *Open Air Metal: Outdoor Metallic Sculpture from the XIXth the Early XXth Century*. Section francaise de l'institut international de conservation (SFIIC).

International Lead Zinc Research Organization, Inc. (1980). *Engineering Properties of Zinc Alloys*. New York.

Johnen, H. J. (2008). "Zinc." In F. Habashi, *Alloys: Preparation, Properties, Applications*. Pp. 121–126. Weinheim, Germany. Wiley-VCH Verlag GmbH.

Kuklik, V, Kudlacek, J. (2016). *Hot-Dip Galvanizing of Steel Structures*. Oxford: Elsevier.

Leygraf, C. Wallinder, I.O., Tidblad, J. Graedel, T. (2016). *Atmospheric Corrosion*. Hoboken, NJ: John Wiley and Sons.

Maass, P., Peissker, P. (2011). *Handbook of Hot-Dip Galvanization*, Weinheim, Germany: Wiley VCH.

Mital, J. (2011). *Bidri Ware and Damascene Work*. Jagdish and Kamla Mital Museum of Indian Art, Hyderabad.

References

Porter, Frank (1991). *Zinc Handbook: Properties, Processing, and Use in Design.* New York: Marcel Dekker, Inc.

Revie, R. W., Uhlig, H. H. (2008). *Corrosion and Corrosion Control.* Hoboken, NJ: John Wiley and Sons.

Rheinzink (2011). *Applications in Architecture.* Datteln, Germany.

Zhang, Xiaoge G. (1996). *Corrosion and Electrochemistry of Zinc.* New York: Plenum Press.

Index

A

abrasives, 71, 221
   fabrication and, 174–175, 177
   maintenance and, 263, 275–277, 287–288
AC. *See* alternating current
acetic acid, 143, 236, 265, 275, 283
acetone, 262, 270
adhesives, 189, 199
   cleanliness and, 272–274, 276–277
Alloy Z15006, 45–47
Alloy Z21220, 47
Alloy Z35631 (ZA 12), 54, 60–61, 211
Alloy Z35636 (ZA8), 54, 60–61, 211
Alloy Z35841 (ZA27), 53–54, 60–61, 193–194, 211
Alloy Z41121, 49–50, 80
Alltrista Zinc Products Company, 17
alternating current (AC), 104, 197, 294
aluminum, 44–45
   anodizing, 104
   coating, 150, 217
   density of, 213
   linear expansion of, 202
   machining and, 195
   melting point of, 168
   mining of, 19
   oxide layers and, 220, 224, 240, 242
   shear strength of, 177
   thermal conductivity of, 196
   zinc-aluminum-magnesium, 142, 211, 219–220
Aluzinc, 220
American Civil War, 15, 18–19
American Galvanizers Association, 297
American Society for Testing Materials, 229
ammonium chloride, 39–40, 139, 223, 283, 289
ammonium hydroxide, 104, 269–270, 277, 283

Anatolia, 2
Ando, Tadeo, 100–101
angel hair finish, 66, 72, 132
anisotropic properties, 34, 47, 49, 52–53
   fabrication and, 186–187, 214
   of rolled sheet, 131, 133–134
anodes, 17, 39, 47, 160
   corrosion and, 222, 225, 239–240, 244–247
anodizing, 104
Anthrazinc™, 35
anti-inflammation cream, 28
antimony, 96, 197, 298
Architectural Metal Series, 41
Architectural Sheet Metal (SMACNA), 27
Architectural Sheet Metal Manual, 26
architecture, 149
   rolled forms, 47–53
   zinc alloy in, 24–28, 41–44
areas, ratio of, 240, 242–245
argon, 197
arsenic, 29
art, 18–24
Art Gallery of Alberta, 89–91
ARTAZN, 17
artifacts, 249–251
Asia, 17
atom, zinc, 9–10
Australia, 17
Austria, 15

B

back-side corrosion, 116, 188
Bahamas, 233, 286
baking soda, 275
Ball Corporation, 16–17, 41
Ball Metal and Chemical, 27
bar tops, 71, 105, 109, 251, 258, 288

Barnard College, New York, 125, 133
Baroque™ patina, 66, 83, 126, 132, 183, 258
batch galvanizing, 138–142, 220, 297
battery, 39–40, 47
   corrosion and, 222–224
beech wood, 255
Belgium, 15, 166
bend test, 199
bending, 184–185, 187
benzyl alcohol, 276
Berlin, 14, 24, 148
Bethlehem Steel, 219
Bidriware, 13
billets, 44, 160
bimetallic corrosion, 234, 239, 245–246
birch wood, 237
bird waste, 235, 263, 267, 271, 277
bitumen, 287, 289
blackened zinc, 100
   corrosion and, 203
   expectations, 121–122, 131
   finish and, 66, 77–79, 84–85
   maintenance of, 258
blanking, 177
boiling point, of zinc, 3
bolting, 204–205
boric acid, 142
Boy and a Fish statue, 249–251
brake forming, 187–191
brass, 2
Bridgeport, Connecticut, 22
Bristol Brass Company, 11
bronze sculpture, 19, 193, 249
Brussels, 24
bumping, 180–184
butyl acetate, 273–274

C

cadmium, 3, 29, 33, 79, 228
   impurity and, 36, 40, 43
   zinc alloys and, 43–45, 47, 54, 59–60
calamine, 1–2, 12, 17, 117, 225. *See also* sphalerite
calcium chloride, 285–286
calcium magnesium carbonate, 89
calots, 39

came edging, 194
Canada, 53, 55, 150
capacitor discharge (CD) stud welding, 199–200
carbide-tipped tooling, 195
carbon dioxide, 2, 11–12, 120
   corrosion and, 216, 226, 228, 232, 237
   finishes and, 65–66, 79, 90, 92
   maintenance and, 257, 259, 281, 286
carburetors, 38, 108, 288
casting, 9, 208–209
   die casting, 167–168, 210–211
   Dough Boy sculptures, 19–20, 22, 166–167
   gravity, 59–61
   melting point and, 33–34, 38, 55–56, 110
   plaster mold, 210, 212
   sand, 20, 56, 169–171, 210, 212
   slush, 20, 34, 56–58, 167, 210–212
   spin, 212–213
   weight comparisons of, 213
   zinc alloys and, 55–56
cathodes, 17, 218, 240, 243, 246
CD. *See* capacitor discharge
cedar wood, 237
Center for The Arts, New Mexico State University, 207–208
centrifugal casting, 208
Cesena, FiorenzoValbonesi, 24, 26, 182
Champion, William, 11, 13
Chattanooga, Tennessee, 87–89, 130
chemical cleanliness. *See* cleanliness
chestnut wood, 255
Children's Hospital of Richmond Pavilion, 85
China, 2, 11–12
chlorides, 79, 84, 115–116
   corrosion and, 225, 238, 249, 252–253
chlorinated rubber, 171
chromium, 55, 69
   mining of, 19
Civil War, US, 15, 18–19
cleanliness
   chemical
      coastal exposure, 286–287
      construction dust, 287–288
      deicing salts, 284–286

hard water deposits, 279–280, 289
  laser ablation and, 289–291
  rust stains, 279–280, 288–289
  storage staining, 281–284
galvanized steel and, 296–299
glass cleaners, 108, 269–270, 277
isopropyl alcohol, 267, 269–273, 277, 293
mechanical
  creep and, 295–296
  dent repair, 294, 296
  distortions, 294–295
  hail damage and, 294
  heat distortion, 294, 296
  scratches and, 292–293
mineral spirits, 270, 272–273, 287, 289
natural zinc and, 257, 268–269, 278, 293
physical
  adhesives, 272–274, 277
  concrete, 276–277
  dirt, 271–272, 277
  fingerprints, 267–271, 277
  graffiti, 275–277
  grease deposits, 274–275, 277
  grime, 271–272, 277
  inks, 276–277
  mill oils, 268, 277
  mortar, 276–277
  sealant decomposition, 273–274, 277
tools for, 288
  deionized water, 278
  high-pressure washing, 277
  laser ablation, 289–291
  prevention, 291
  steam cleaning, 278
clear coating, 71, 77, 100, 119, 133
  maintenance, 248, 269
  of sheet zinc, 157–158
cleats, 118–119, 133–134, 205
CNC. *See* computer numeric control
coastal environment, 75, 77, 115–116
  corrosion and, 225, 229–230, 233
  maintenance and, 285–287, 297
coil-coated zinc, 8

coiled zinc, 153–156
  galvanized, 138–141
cold flow, 106, 205, 248
cold working, 9, 31, 34, 40, 150, 248
  fabrication and, 158, 184
cold-forming, 294–295
color matching, 122–124
Cominco Ltd., 53, 150
computer numeric control (CNC), 178, 180, 197
concrete, 276–277
condensation, 68, 115–116
  corrosion and, 240, 245, 247
  exposed surfaces and, 230–231, 235–236
  maintenance and, 261, 266, 268
construction dust, 287–288
continuous casting, 36, 41, 122, 150–151, 184, 215
continuous galvanizing, 138–141, 162
copper, 2, 65
  density of, 213
  linear expansion of, 202
  melting point of, 148
  mining of, 19
  plated sculpture, 11, 19–23, 84, 166, 193, 195
  in rolled zinc, 48–52
  shear strength of, 177
  soldering and, 195
  thermal conductivity of, 196
  titanium alloy, 40–41, 45, 48–53
    creep and, 136–137, 295
    in fabrication, 184, 193
    finishes, 105–107, 113, 136–137
  US penny, 17, 34, 39, 53, 84, 196
  voltage potential for, 243
copper nitrate, 265
corner folds, 191
corrosion, 8. *See also* galvanic corrosion
  of artifacts, 249–251
  battery and, 222–224
  bimetallic, 234, 239, 245–246
  blackened zinc and, 203
  carbon dioxide and, 216, 226, 228, 232, 237
  chlorides, 225, 238, 249, 252–253
  coastal environment, 225, 229–230, 233
  condensation and, 240, 245, 247

corrosion (*continued*)
  deicing salts, 251–252
  electrolyte effects, 244–245
  exterior exposure and, 228–235
  fertilizers, 253–254
  fretting, 234
  of galvanized steel, 218–219
  geometric relationships and, 243–244
  intergranular, 234, 248
  interior exposure and, 227–228
  introduction to, 215
  organic corrosives, 30, 254–255
  oxygen and, 217, 235–236
  pitting, 234, 246–247
  protective zinc coating, 216–218
  rates of, 229
  resistance, 31, 38
    preweathering and, 74–75, 77
    zinc phosphate and, 101–102
  saponification, 254–255
  sheltered exterior, 230–235
  sherardizing, 221–222
  of statues, 249–251
  stress cracking, 248–249
  thermal spray and, 221–222
  of underside surfaces, 236–237
  uniform, 235–236
  urban environment and, 229–230, 249–250
  white corrosion, 237–239, 281–282, 289
  zinc carbonate and, 216, 218, 225–227, 230
  zinc hydroxide and, 225–226, 229, 231–232, 235
  zinc powder, 220
corrosion products, 116–117
corrugation, 137, 140, 180
  galvanized sheet, 206–207, 223–224, 240–241
counterfeht, 1, 3
countertops, 109, 288
Cowpers-Coles, Sherard, 221
cracks, in sculpture, 24, 84, 111–112
  from creep, 169–171, 295–296
  stress corrosion, 248–249
  temperature and, 187
creep, 24
  corrosion cracking, 248–249
  dimensional changes, 295–296

  expectations and, 106–107, 111–112, 135–137
  grain and, 137
  mechanical cleanliness, 295–296
  resistance to, 16, 40–41
  sand cast and, 169–171
  sculpture and, 169–171, 295–296
  stages of, 136
  titanium and, 44, 136–137, 295
  wrought zinc and, 45, 47
crevice corrosion, 234
crystal structure, 3
  dendrites and, 5–7, 96
Curie, Marie, 63
custom patinas, 79–80, 124–131
cutting, 177
  laser, 178
  plasma, 179
  waterjet, 179–180

D

dark variegated patinas, 80–87
darkened galvanized steel, 100–101, 142–143.
        See also blackened zinc
Davis, Miles, 173
dealloying, 234
deicing salts, 108, 215, 235, 251–252
  cleanliness and, 260–261, 284–286
deionized water, 245, 262, 267–270
  pressure washing, 263, 277–278
Delach, Austria, 15
dendrites, 6–7, 96
dents, 258, 262
  repair, 294, 296
Denver, Colorado, 111–112
diammonium phosphate, 253
die casting, 33, 58–59, 167–168, 210, 210–211
Diesel Building, Chicago, 143
Diller Scofidio and Renfro, 126
dimensional change, 187, 295–296
dirt, cleaning of, 271–272, 277
dispersion hardening, 136–137
distilled water, 245, 263, 278
distortions, 294–295
dolomite, 87, 89
Dony, Jean-Jacques, 11–13, 16

Dough Boy sculpture, 19–20, 22, 166–167
Douglas fir, 189, 237, 255
downpipes, 117–118, 247
dry-cell battery, 39–40

E
Egyptians, 12
elasticity, 8, 135–136, 193, 214
electrical conductivity, 8, 197, 214
electrochemical potential, 216–217
electrolyte effects, 244–245
electrolytic method, 17
electroplating, 19, 96, 103, 141
electro-potential. *See* voltage potential
electrowinning, 17, 30
element 30, 8–9, 31. *See also* zinc
elongation, in rolled zinc, 49–53
embossing, 107, 164
England, 11, 13
environmental effects
   coastal, 74, 77, 115–116
      corrosion and, 225, 229–230, 233
   maintenance and, 266–267, 285–287, 297
   rural, 229, 266, 297
   urban, 74, 117, 121, 128, 259
      corrosion and, 229–230, 249–250
epoxy, 171, 220
etching, 9
ethanol, 38, 272
ethyl acetate, 272–274
Europe, 14, 150, 187, 189, 193, 208, 227, 288
   ornamentation and, 165–168
eutectic mixture, 136–137
exterior corrosion
   exposure and, 228–230
   sheltered surfaces, 230–235
extruded zinc alloys, 53–55
extrusion, 53–54, 158–159, 194

F
fabric, of zinc, 102–103
fabrication
   anisotropic properties, 185–187
   blanking, 177
   bolting, 204–205
   brake forming, 187–191
   bumping, 180–184
   casting, 208–209
      die casting, 210–211
      plaster mold, 210, 212
      sand, 169–171, 210, 212
      slush casting, 210–212
   cutting, 177
      laser, 178
      plasma, 179
      waterjet, 179–180
   extrusion, 53–54, 158–159, 194
   fastening, 204–205
   forging, 193–194
   forming, 184–185
   grain direction and, 185–187
   hot-dipped galvanizing, 175, 206–208
   machining, 194–196
   perforating, 180–184
   punching, 180–184
   roll forming, 191–192
   saw cutting, 177–178
   shearing, 177
   of shingles, 187–189
   soldering, 195–197
   springback, 187
   storage and handling, 174–176
   superplastic forming, 192–193
   thermal spray, 206
   v-cutting, 191
   welding, 197–201
   zinc attributes, 173–174
Faraday's laws, 246
fastening, 204–205
ferric chloride, 142
fertilizers, 253–254
Filix Nussbaum Haus, 24
fingerprints, 259
   cleanliness and, 267–271, 277
finishes, 8
   angel hair, 66, 72, 132
   anodizing, 104
   appearance and, 65–68
   blackened zinc, 66, 77–79, 84–85

finishes (*continued*)
  carbon dioxide and, 65–66, 79, 90, 92
  clear coating, 71, 77, 100, 119, 133
    of sheet zinc, 157–158
  color matching, 122–124
  flatness and, 131–135, 153
  galvanized steel, 93–100, 138, 142–143
  glass bead, 66, 71–72, 132
  grain and, 149–151, 155
  introduction to, 63–65
  mechanical finishes, 71–73
  mill produced, 68
  natural, 63–64, 66–67
    on cast surface, 110–112
    on sheet, 107–110, 174
  of Paris roofs, 69, 166
  patina
    custom, 79–80, 124–131
    dark variegated, 80–87
    iridescent, 93
    zinc oxide, 87–93
  preweathering, 73–77, 113–119
  relative gloss, 131–132
  sherardizing, 103, 141, 221–222
  statues and, 84–85
  thermal spray, 103, 141, 160, 171, 206
    corrosion and, 221–222
  titanium and, 105–107, 113, 136–137
  visual distortion and, 131–135
  zinc fabric, 102–103
  zinc phosphate, 101–102
  zinc-copper-titanium alloy, 105–107, 136
FiorenzoValbonesi Cesena, 24, 26, 132
fir wood, 189, 237, 255
Fiske Company, J.W., 23
flatness, 131–135, 153
fluoride, 104, 143, 278
fluxing material, 139, 146
  soldering and, 195–197
foam insulation, 236
foil, 158
Folly Theater, Missouri, 24, 26
Food and Drug Administration, U.S., 28

forging. *See also* casting
  processes of, 193–194
  zinc alloys, 53–55
formaldehyde, 189
forming, 184–185
  brake forming, 187–191
  cold-forming, 294–295
  roll forming, 191–192
  superplastic, 31, 192–193
  temperature and, 187
foundries, 22, 24, 34, 59–60, 110
  casting and, 169–170, 208
France, 15, 18, 63, 208. *See also* Paris, France
fretting corrosion, 234
fumes, of zinc oxide, 8, 29, 169, 179
fusion stud welding, 199–200

G
galena, 2
galfan, 142, 219–220
galvalume, 142, 150, 219–220
Galvani, Luigi, 15
galvanic cell, 240, 245–247
galvanic corrosion, 204, 234, 239–247, 266
  determining factors, 242
  distance and, 244
  temperature and, 245–246
galvanized steel, 218–219
  cleanliness, 296–299
  darkened, 100–101, 142–143
  galvanic corrosion, 204, 234, 239–247, 266
  roll forming, 191–192
  spangle and, 5–6, 95–98, 138
  structural shapes, 99–100
  surface of, 93–99
  zinc carbonate and, 140, 296, 299
  zinc hydroxide and, 97, 100, 108, 140–143
  zinc phosphate and, 101–102
galvanizing, 25
  batch galvanizing, 138–142, 220, 297
  continuous galvanizing, 138–141, 162
  electrochemical potential and, 216–217
  facility for, 7, 16, 98
  recycling and, 29–30
gamma rays, 3

Index 313

Gandhi, Mahatma, 257
gas tungsten arc welding (GTAW), 197, 294
Gayle, Margot, 27
Germany, 11, 14–15, 24, 40, 63, 150, 208
Glacier Gray™, 35
glass, 194
    bead finish, 66, 71–72, 132
    cleaners, 108, 269–270, 277
    corrosion and, 255
    linear expansion of, 202
Goddess of Liberty statue, 21, 23–24
gold, 1, 9, 13, 34, 65, 217
gold leaf, 254–255
Goslar, Germany, 11
graffiti, 275–277
grain
    alignment, 44, 48–53, 75, 96, 133–134
    creep and, 137
    direction, 185–187
    finishes and, 149–151, 155
gravity cast alloys, 59–61
grease deposits, 274–275, 277
grime, cleaning of, 271–272, 277
Grissom, Carol, 193
GTAW. See gas tungsten arc welding
gunmetal, 19
gutters, 40, 117–118

H

hail damage, 294
hammer tone texture, 163
hard water deposits, 279–281, 289
Hausmann, Baron, 14
Hazelett Machine, 41
health, 28–30
heat distortion, 294, 296
Hegeler, Edward C., 16
helium, 197
Helix Architecture, 78, 181
Hemimorphite, 16
herringbone pattern, 188
High Grade (HG) alloy, 36–37
high-pressure washing, 277, 287
Hoffman, Reilly, 105–106
Holtzman, Malcolm, 206

horizontal retort, 11–12, 16
hot-dipped galvanized, 106, 138–142
    corrosion, 219–224
    fabrication, 175, 206–208
    maintenance, 266, 297–298
    process of, 37, 94–99
humidity, 68, 124–125, 228
Hunter Museum of Art, Chattanooga, 87–89
Hunter™ patina, 48, 66, 87, 89–91, 238
    fabrication, 189–190
    maintenance, 258, 283
    on sheet zinc, 153–154
    weathering, 118, 129–130
hydrochloric acid, 142, 277
hydrogen peroxide, 275
hydrozincite, 16
hypochlorous acid, 252–253

I

Illinois, US, 15–16, 48
ILZRO. See International Lead Zinc Research Organization
ImageWall®, 181
immune system, 28–29
Independence, Missouri, 165
India, 2, 11–12, 13
industrial atmosphere, 74, 120–121, 297
    corrosion and, 225, 229–230, 233, 235
infrared light, 9, 133, 203
ingot alloys, 36–39
inhalation hazard, 8, 29, 169, 179
inks, 276–277
Institute for Contemporary Art, Virginia, 74, 76, 134–135
intergranular corrosion, 234, 248
International Lead Zinc Research Organization (ILZRO), 60
ion exchange resins, 278
iridescent patina, 93
iron, 23
    alloys, 43–45, 54, 59–60
    corrosion and, 239
    density of, 213
    galvanizing, 34, 37, 139–140, 142, 216
    health and, 28

iron (*continued*)
  impurity and, 40
  melting point of, 148
  mining of, 19
  patinas and, 79, 81, 94–95, 98
  zinc peel and, 225
iron oxide, 81, 278–280, 289
isocyanate adhesives, 189
isopropyl alcohol, 267, 269–273, 277, 293

J
jack, 17, 29
Jarden Corporation, 17
Jernigan, Tex, 125–126
Jewish Museum, Berlin, 24
J.L. Mott Iron Works Company, 23
Joplin, Missouri, 15–17, 17
Judd, Donald, 87–88, 269
J.W. Fiske Company, 23

K
Kansas, US, 15
Kansas City, Missouri, 24, 26
  Plaza statue, 79
Kelmis, France, 14
kerf, 178–179
kirksite, 34, 60, 62–63
Korloy™, 53–54, 193–194

L
La Vielle Montagne Zin Mining Company, 14
lacquer thinners, 262
lanaphilosophica, 3
LaSalle Zinc Works, 16, 27–28
lasers, 178
  ablation, 283, 289–291
lathes, 197
lead, 43–44
  impurity, 36, 40
  melting point of, 148
  tin solders, 146, 195
Libskind, Daniel, 24
Liege, Belgium, 15
Look, David, 27
Louisiana, US, 15

M
machining, 194–196
magnesium, 19, 44–45, 186
  casting alloy, 59–60, 168
  coatings and, 216–217, 219–220
  forging alloy, 54
  mining of, 19
  in seawater, 217, 287
  in water, 278, 280
  zinc-aluminum alloy, 142, 211, 219–220
magnesium chloride, 285–286
maintenance. *See also* cleanliness
  abrasives and, 263, 275–277, 287–288
  carbon dioxide and, 257, 259, 281, 286
  of clear coating, 248, 269
  coastal environment, 285–287, 297
  condensation and, 261, 266, 268
  environmental effects, 266–267, 285–287, 297
  introduction to, 257–258
  preweathered and, 258–259, 264–266, 277
  schedule for, 130, 260–262, 260–264
  zinc surfaces and, 258–259
manganese oxide, 39–40, 223
mansard roofs, 24, 69, 148–149
marine environment. *See* coastal environment
Maris, Roger, 1
Marmatite, 16
mars, cleaning of, 292–293
Matthiessen, Fredrick W., 16, 27
Matthiessen and Hegeler Zinc Co., 41, 48
Max Brenner™ chocolate company, 83, 125–126
Mazak, 34, 168, 210
mechanical finishes, 71–73
MEK. *See* methyl ethyl ketone
melting point, 8–9
  of aluminum, 146, 148, 168
  casting and, 33–34, 38, 55–56, 110
  of copper, 11, 146, 148
  of iron, 148
  lasers and, 178
  of lead, 146, 148
  sand casting and, 56, 169
  sherardizing and, 221
  slush casting and, 167, 210–211

spangle and, 96
thermal expansion and, 202
of tin, 146, 148
welding and, 197
of zinc, 145–146, 148
mercury, 3
Mesker Brothers, 26–27
metallizing, 99, 160, 206
Metals in America's Historic Buildings (Gayle, M. and Look, D.), 27
methanol, 272, 275
methyl ethyl ketone (MEK), 272–274, 287, 289
Meyers, Elijah E., 21
Microzinc, 16, 27–28, 35, 41
milling, 9
   finishes, 68
   oils, 268, 277
mineral spirits, 270, 272–273, 287, 289
mining, 11, 19, 29–30
   companies, 14–15, 53, 150
   of lead, 44
mischmetal, 219–220
Missouri, US, 15–17, 26, 29, 165
M.J. Seelig and Company, 23, 208
molten bath, 6, 68, 93–94, 102, 297
molybdenum oxide, 93
monkey metal, 249
monoammonium phosphate, 253
Monumental Bronze Company, 22, 209
Morimoto Restaurant, New York, 100–101
mortar, 276–277
Mott Iron Works Company, J.L., 23
Mullins, W.H., 193
Muntz, George Fredrick, 11
Muntz metal, 11, 177

N

Napoleon, 11, 14, 105
National Association of Sheet Metal Contractors, 26–27
natural zinc, 3, 152
   cleanliness, 257, 268–269, 278, 293
   color, 63–64, 66–71
   corrosion, 236, 251
   fabrication, 174, 182, 197

finish
   on cast surface, 110–112
   on thick sheet, 109–110
   on thin sheet, 107–109
New Jersey Zinc Company, 16, 40, 168, 211
New Mexico State University, 207–208
nickel, 9, 227, 239
   mining of, 19
nickel chloride, 143
nickel silver, 217
noble metal, 216–217, 244–247, 266
North America, 160, 189
   alloys and, 41, 55–56, 150, 210
   architecture and, 26–28, 149
   casting in, 167–168, 210
   finishes and, 63, 65
   health and, 28–29, 150
   history and, 15–16
   lead standard, 36
   rainfall and, 227
Northwest Territories, Canada, 55, 150–151
nucleation, 96
nylon structural matte, 189–190, 236–237

O

oak wood, 236–237, 255
oil canning, 69, 107, 131, 164
Oklahoma, US, 15
optical profile analysis, 127–128
organic corrosives, 30, 254–255. *See also* corrosion
orichalcum, 2
oriented strand board (OSB), 218, 236
ornamentation, 14, 24–25, 169, 295
   casting and, 166–168
   finishes and, 63
   spinning of, 197–198
   stamping of, 165–166
OSB. *See* oriented strand board
Osnabrück, Germany, 24
oxalic acid, 269–270, 277, 279, 289
oxidation, 9, 89, 92, 116, 124, 179
   layer, 220, 224, 240, 242
   mill oils and, 268
   pitting and, 247
   welding and, 197

oxidation-reduction reaction, 234, 241–242, 297
oxygen, 11–12, 17, 29
    atom, 9–10
    corrosion and, 217, 235–236
    finishes and, 65, 79, 89
    galvanic corrosion and, 240–242, 245–246
    maintenance and, 257–258, 273, 297
        chemicals, 279, 281

P
pack rolling, 113, 149–151
packaging, 281, 291–292
packs, 149–150
paint stripper, 275
Palau Sant Jordi, 52
panels
    cast zinc, 56, 61, 111, 150
    spandrel, 193–194
    stamped, 15, 24, 29, 148, 165, 194, 295
    systems, 189–190
Paracelsus, 11
Paris, France
    roofs of, 11, 14, 24–25, 131, 166
        finishes, 69, 166
        maintenance, 258
    skyline of, 11, 63, 148
Parkerizing treatment, 102, 143, 292
patinas, 9. *See also* Hunter™ patina; Roano™ patina
    Baroque™, 66, 83, 126, 132, 183, 258
    custom, 124–131
    dark variegated, 80–87
    shingles, 157–158
    zinc oxide, 87–93
Payne, Rob, 33
Pelzel, Erich, 40–41
Peninsula Hotel, Hong Kong, 158–159
penny, copper, 17, 34, 39, 53, 84, 196
perforated zinc, 162–163, 174, 180–184
periodic table, 3–4
permanent mold casting, 210, 212
pewter, 1, 71, 79
phenol, 189, 255
phosphoric acid, 73, 104, 143, 157, 250
    maintenance and, 265–266, 275, 279–280, 292

physical cleanliness. *See* cleanliness
Pin, Gino, 55, 151
pine wood, 189, 237, 291
piping, 117–118, 247. *See also* tubing
pitting corrosion, 234, 246–247
planter boxes, 110, 153
plants, 29, 253–254, 261
plasma cutting, 179
plaster mold casting, 210, 212
plastic coverings, 175–176, 178, 235, 273
plastic deformation, 34, 106, 136, 182, 192, 294. *See also* creep
plate, zinc, 152–153
    thick plate, 41, 105, 109–110, 113, 205
    thin sheet, 107–109, 141
plating, 9
    of copper penny, 17, 34, 39, 53, 84
    copper plating, 11, 19–23, 84, 166, 193, 195
    electroplating, 19, 96, 103, 141
plywood, 189, 236–237
Poland, 14–15
Polk County Criminal Court Building, Iowa, 87
pollutants, 108, 130, 222–223, 285, 288
polyester, 171, 255
pot metal, 249
potassium, 253, 265, 287
powder, zinc, 171–172
Presbyter, Theophilus, 2
pressure die casting. *See* die casting
pressure washing, 263, 271, 274, 277, 280
Prestal®, 44
preweathering, 73–77, 115–119
    expectations of, 120–121
    exterior surfaces and, 228–235
    maintenance and, 258–259, 264–266, 277
    zinc carbonate and, 113–114, 120–121, 132, 232
Prime Western Grade (PWG), 36
Prime Western Grade recycled (PWG-R), 36
protective wraps, 175–176, 178, 235, 273. *See also* corrosion
Prussia, 14–15, 24
pullover, 204
punching, 180–184
PWG. *See* Prime Western Grade

PWG-R. *See* Prime Western Grade recycled
pyrometallurgic process, 11, 17

R
radiation, 3
Rajasthan, 13
Randall Stout Architects, 85–91
ratio of areas, 240, 242–245
recycling, 9, 29–30
   PWG-R, 36
Red cedar, 237
reduction reaction, 234, 241–242, 297
reflectivity, 7, 65–69, 73, 107, 126, 294, 296
Regis College Campus, Denver, 111–112
relative gloss, 131–132
resistant welding, 201
restoring, 15, 278
   of buildings, 111, 165
   of mechanical strength, 193
   of preweathering, 264–265
RLE. *See* Roast-Leach-Electrowin
Roano™ patina, 35, 66, 231
   gloss of, 132–133
   maintenance, 258, 271–272
   natural tones of, 80–83, 85–87, 91
   on sheet zinc, 180–181
   on shingles, 157–158, 188–189
   weathering of, 124–129
Roanoke, Virginia, 80–83, 85–86, 128–129
Roast-Leach-Electrowin (RLE) process, 17
Rockwell hardness, 46–47, 49–52
rod, 160–161, 222
roll forming, 191–192
rolled zinc, 39–41, 153–158
   anisotropic properties, 131, 133–134
   copper in, 48–52
   elongation and, 49–53
   pack rolling, 113, 149–151
   preweathering and, 73–75
   tensile strength of, 45–52, 134
   textures, 72–73
   titanium in, 44–45, 47–49, 52
Romans, 1–2, 12
roofs, 45, 247
   mansard, 24, 69, 148–149

   of Paris, 11, 14, 24–25, 131, 166
      finishes, 69, 166
      maintenance, 258
rosin paper, 189
roughness, 6, 109, 127
rubber
   coatings, 171
   corrosion and, 255
   linear expansion of, 202
rural environment, 229, 266, 297
rust stains, 279–280, 288–289

S
sand casting, 20, 56, 169–171, 210, 212
saponification, 8, 254–255
satin finish, 65, 71, 292
saw cutting, 177–178
SchedulaDiversariumArtium (Presbyter, T.), 2
scheduled maintenance, 260–264
Schinkel, Karl Friedrich, 14, 24
Scotch pine, 237
ScotchBrite™, 143
scratches, 292–293
sculpture
   bronze, 19, 193, 249
   copper plating, 11, 19–23, 84, 166, 193, 195
   cracks in, 24, 84, 111–112
   creep and, 169–171, 295–296
   deicing salts and, 251–252
   Dough Boy, 19–20, 22, 166–167
   maintenance log, 260–264
   stress cracking, 248–249
sealants, 287, 289, 295
   decomposition of, 273–274, 277
seawater, 216–217, 243–245, 287
Seelig, Moritz J., 23, 208
self-annealing, 169, 184
semi-rigid foam insulation, 236
shear strength, 177, 214
shearing, 177
Sheet Metal Air Conditioning National
     Association (SMACNA), 27
Sheet Metal Publication Company, 27
sheet zinc, 41, 43, 72–73, 236
   natural finish, 107–109, 174
   rolled, 153–158

sheet zinc (*continued*)
   textured, 163–164
   thin sheet, 141, 178, 184–188
   Type 1, 46, 49
   Type 2, 46, 51
sheltered exterior corrosion, 230–235
sherardizing, 103, 141, 221–222
shingles, 83, 236–237
   fabrication of, 187–189
   patina and, 157–158
shipping, 73, 137, 291
Silesia, Poland, 14–15
silicate-based coatings, 171, 220, 262
silicone, 199, 212, 273–274, 277
silver, 1, 3, 9, 13, 34, 67
   corrosion and, 217, 227–228
simonkolleite, 233
skin ointment, 28
Slate™, 35
slush casting, 20, 34, 56–58, 167, 210–212
SMACNA. *See* Sheet Metal Air Conditioning National Association
smelting, 29
smithsonite, 16–17
sodium acetate, 142
sodium bicarbonate, 275
sodium carbonate, 73, 262
sodium chloride, 285–286
sodium fluoride, 143
sodium metasilicate, 262
sodium phosphate, 256
sodium pyrophosphate, 262
sodium silicate, 256
soldering, 9, 22, 29, 146, 195–197
   distortion and, 294–296
SOM Architects, 125
Sorel, Stanislas, 15
spandrel panel, 193–194
spangle, 70
   galvanized steel and, 5–6, 95–98, 138
Special High Grade (SHG) alloy, 36
Specific gravity, 8
spelter. *See* zinc
Sphalerite, 16

spin casting, 212–213
springback, 187
spruce wood, 237
St Charles Hotel, New Orleans, 15
St. Louis, Missouri, 26
stained-glass windows, 194
stainless steel
   density of, 213
   linear expansion of, 202
   melting point of, 146
   shear strength of, 177
   thermal conductivity of, 196
stains
   by fertilizers, 253–254
   hard-water deposits, 279–281
   rust, 279–280, 288–289
   from storage, 281–284, 291
   wet storage, 237–239, 283
   white storage, 174–175, 238, 281–282, 289
stamped forms, 166, 193
stamped panel, 15, 24, 29, 148, 165, 194, 295
Standard Practice in Sheet Metalwork, 27
Stanford McMurtry Building, 126
Starck, Philippe, 159
Statue of Liberty, 57, 211
statues. *See also* sculpture
   bird waste and, 235, 263, 267, 271, 277
   Boy and a Fish, 249–251
   corrosion and, 249–251
   deicing salts, 251–252
   finishes, 84–85
   Goddess of Liberty, 21, 23–24
   Kansas City Plaza, 79
la statuomanie, 18
steam cleaning, 263, 275, 278
steel. *See also* galvanized steel
   electrochemical potential, 216–217
   linear expansion of, 202
   metallizing of, 99, 160, 206
   shear strength of, 177
   stainless, 146, 177, 196, 202, 213
   structural shapes, 99–100
   thermal conductivity of, 196
   weathered, 92, 146
   zinc coatings and, 219–220

Index

steel mesh, 206, 208
stelleite, 81
Steven Holl Architects, 74, 76, 134–135
Stolberger Zinc AG, 40–41
storage, 174–176
   staining and, 278, 281–284, 291
   wet stains, 237–239, 283
   white corrosion, 237–239, 281–282, 289
Stout, Randall, 85–91
stress corrosion cracking, 234, 248–249
stud welding, 199–200
subdendrites, 6
sulfates, 225, 278, 280
sulfur dioxide, 17, 121, 230
super plasticity, 31, 44, 192–193
Symposium on Atmospheric Corrosion of Non-Ferrous Metals, 229

T

Taubman Museum of Art, 80–83, 85–86, 128–129
Taylor Architectural Group, 55
temperature
   corrosion and, 245–246
   in fabrication, 184–185, 187
   heat distortion, 294
   metal surface and, 284
   wrought forms and, 174
Tennessee, US, 17, 87–88, 130
tensile strength, 41, 44–47, 49–50
   bend test and, 199
   of cast alloys, 59–60
   of die-cast alloys, 168
   fabrication and, 146, 214
   finishes and, 107
   of forging alloys, 54
   of kirksite, 62
   of rolled zinc, 45–52, 134
Texas, US, State Capital Building, 21, 23
texture, 164
   hammer tone, 163
   mechanically rolled, 72–73
   rolling, 107–108
thermal conductivity, 8, 195–196, 214
thermal expansion, 69, 111, 131–134, 202, 202–204, 214

thermal spray, 103, 141, 160, 171, 206
   corrosion and, 221–222
thick plate, 41, 105
   bolting of, 205
   natural finish on, 109–110
   preweathered, 113
thin sheet, 141, 178
   forming, 184–188
   maintenance, 259, 294
   natural finish on, 107–109, 174
TIG. *See* tungsten inert gas
tin
   coatings, 217
   lead solders, 146, 195
   melting point of, 146, 148
Titanaloy, 16, 27, 41, 48
titanium, 16
   corrosion and, 215, 217
   creep and, 44, 136–137, 295
   extrusion and, 53–54
   finishes and, 105–107, 113, 136–137
   linear expansion of, 202
   oxide layers, 240, 242
   preweathereing, 113
   in rolled zinc, 44–45, 47–49, 52
   sheets and, 153, 184–185
   thermal conductivity of, 196
   thermal expansion of, 196, 202
   zinc-copper alloy, 25, 40–41, 45, 47–53
      corrosion, 229
      fabrication, 150, 184, 193
toluene, 262, 273, 275
tools, for cleaning, 288
   deionized water and, 278
   high-pressure washing, 277
   laser ablation, 289–291
   prevention, 291
   steam cleaning, 278
toxicity, 28–29
   zinc fever, 8, 169, 179
tubing, 110, 145, 159
tumbling, 103, 221
tungsten inert gas (TIG) welding, 197–199, 294, 296

Twain, Mark, 145
Type 1 zinc sheet, 46, 49
Type 2 zinc sheet, 46, 51

U
ultraviolet radiation (UV), 9, 28, 39, 77, 119, 158
underside corrosion, 236–237
Unified Numbering System (UNS), 35–36, 41, 43, 49–51, 59
uniform corrosion, 234–236
United Nations Education, Scientific and Cultural Organization (UNESCO), 14
United States (US), 59, 84
    architecture in, 24–28
    casting in, 167, 208
    foundries in, 34
    history and, 15–17
    plywood use, 236
    sculpture replication, 19–20, 23–24, 167
    sulfur dioxide and, 121
The Universal Sheet Metal Pattern Cutter, 27
UNS. *See* Unified Numbering System
urban environment, 74, 117, 121, 128, 259
    corrosion and, 229–230, 249–250
urea, 236
urea-formaldehyde, 189
US. *See* United States
UV. *See* ultraviolet radiation

V
Vaile Mansion, 20, 165
Valbonesi, Fiorenzo, 24, 26, 132
vandalism, 264–265
variegated patinas, 80–87
v-cutting, 191
venting, 188, 274, 276, 289
vibration dampening, 146
Victorian era, 18–20
Vieille Montagne Zinc Company, 15
vinegar, 143, 250, 288
Virginia, US, 80–83, 85–86, 128–129
    Commonwealth University, 74, 134–135
    Institute for Contemporary Art, 74, 76, 134–135

visual distortion, 131–135
VM Zinc, 14
volatile organic compounds (VOCs), 30, 275–276
Volta, Alessandro, 15, 222
voltage potential, 216–217, 243

W
water
    deionized, 245, 262, 267–270
    distilled, 245, 263, 278
    hard water deposits, 279–281, 289
    pressure washing, 263, 277–278
    seawater, 216–217, 243–245, 287
waterjet cutting, 179–180
weathered steel, 92, 146
welding, 9
    CD stud welding, 199–200
    GTAW, 197
    spot welding, 201
    TIG, 197–199
wet storage stain, 237–239, 283
white bronze, 22, 84, 147, 209
white cedar, 237
white rust corrosion, 237–239, 281–282, 289
white storage stain, 174–175, 238, 281–282, 289
Wickerson, Michael, 213
Willemite, 16
wine, 105, 109, 258, 288
Winery Cantina de Il Bruciato, 26, 182
wire, 160
    battery and, 222
    mesh, 161
    thermal spray and, 206, 221
    welding and, 197, 199
    zinc powder and, 171
Wisconsin, US, 15
wood, 255, 291
    plywood, 189, 236–237
World Heritage site, 14
World War I, 19, 24, 34, 84
World War II, 15, 58
woven structural matte, 189–190, 236–237
wrought zinc, 148–152, 174
    alloys, 44–47, 108–109

X

xylene, 262, 270, 273, 275

Y

YAG. *See* yttrium aluminum garnet
Yates, William Butler, 215
yttrium aluminum garnet (YAG), 289–290

Z

ZA alloys
  Z12 (Z35631), 54, 60–61, 211
  Z35841 (ZA27), 53–54, 60–61, 193–194, 211
  ZA8 (Z35636), 54, 60–61, 211
Zahner Engineering, 64, 123
Zahner Metal Conservation, 35, 79, 84, 264, 280
Zamac, 34
Zamak series, 34
  Zamak 3 (Z33520), 43, 59, 168, 210–211
Zawar, Rajasthan, 13
zinc. *See also* corrosion; fabrication; finishes
  as architectural metal, 24–28
  in art, 18–24
  atom of, 9–10
  boiling point of, 3
  carbon battery, 223
  crystal structure of, 3, 5–7
  dendrites and, 6–7, 96
  density of, 213
  deposits, 15
  elasticity of, 8, 135–136, 193, 214
  forms of
    coiled, 153–156
    extrusion, 158–159
    foil, 158
    introduction, 145–148
    plate, 152–153
    powder, 171–172
    rod, 160–161, 222
    sheet, 153–158
    textured zinc sheet, 163–164
    wire, 160–161, 171
    wrought, 148–152
  health and, 28–30
  history of, 11–17
  linear expansion of, 202–204
  lozenges, 28
  mechanical properties of, 214
  melting point of, 96, 98, 145–146, 148
  mining of, 11, 14–15, 29–30, 53, 150
  natural color, 68–71
  ores of, 17–18
  oxidation and, 9–10
  as protective coating, 216–218
  shear strength, 177, 214
  thermal conductivity, 8, 195–196, 214
  tubing, 110, 145, 159
  uses of, 37–38
zinc alloys. *See also* casting
  alloying descriptions, 34–36
  aluminum and, 168, 234, 248
  architectural rolled forms, 47–53
  in architecture, 24–28, 41–44
  casting of, 55–56
  copper-titanium, 25, 45, 47–53, 150
    corrosion, 229
    finishes, 105–107, 113, 136–137
  die casting, 58–59
  extruded, 53–55
  forged, 53–55
  gravity casting, 59–61
  ingots grades, 36–39
  introduction to, 33–34
  kirksite, 60, 62–63
  magnesium-aluminum, 142, 211, 219–220
  rolled forms, 39–43
  slush casting, 56–58
  wrought zinc, 44–47
zinc blende, 2, 17
zinc carbonate, 108, 157
  corrosion and, 216, 218, 225–227, 230
  galvanized steel and, 140, 296, 299
  mild, 235–237
  natural finish and, 110–112
  preweathering and, 113–114, 120–121, 132, 232
    maintenance, 258–259, 264, 266
  rust stains and, 279, 281
  salts and, 285, 287
zinc fabric, 102–103
zinc fever, 8, 29, 169, 179

zinc fever (*continued*)
zinc hydroxide, 70, 77–80
   chemical cleanliness, 281–284, 289, 296
   corrosion and, 225–226, 229, 231–232, 235
   custom patina and, 126, 129
   explosiveness, 172
   galvanized steel and, 97, 100, 108, 140–143
   mill oils and, 268
   oxide patinas and, 87–90
   statues and, 249, 251
   weathering and, 92, 115–116, 121
   wet storage and, 174, 237–238
   zinc coating and, 216–219
zinc hydroxychloride, 115–116, 233
zinc nitrate, 143
zinc oxide
   fumes of, 8, 29, 169, 179
   layers and, 220, 224, 240, 242
   patinas, 87–93
   saponification, 254–255
   skin ointment, 28
zinc peel, 225
zinc pest, 249
zinc phosphate, 71, 73–74, 77
   darkening and, 121, 143
   on galvanized steel, 101–102
   preweathering and, 114, 232, 258–259, 266, 277
   scratches and, 292
zinc powder, 171–172, 220
Zinc Sculpture in America 1850-1950 (Grissom, C.), 193
zinc sulfide, 1–2, 17, 117, 225
Zincalume, 220
zincates, 227
zincite, 16–17